# Functional and Bioactive Properties of Food

# Functional and Bioactive Properties of Food

Special Issue Editors

**Diego A. Moreno**
**Nebojsa Ilic**

MDPI • Basel • Beijing • Wuhan • Barcelona • Belgrade

MDPI

*Special Issue Editors*

Diego A. Moreno
National Council for Scientfic Research
(CEBAS-CSIC)
Spain

Nebojsa Ilic
Institute for Food Technology–Novi Sad
(FINS)
Serbia

*Editorial Office*
MDPI
St. Alban-Anlage 66
4052 Basel, Switzerland

This is a reprint of articles from the Special Issue published online in the open access journal *Foods* (ISSN 2304-8158) from 2017 to 2018 (available at: https://www.mdpi.com/journal/foods/special_issues/Functional_Bioactive_Properties_Food)

For citation purposes, cite each article independently as indicated on the article page online and as indicated below:

LastName, A.A.; LastName, B.B.; LastName, C.C. Article Title. *Journal Name* **Year**, *Article Number*, Page Range.

**ISBN 978-3-03897-354-6 (Pbk)**
**ISBN 978-3-03897-355-3 (PDF)**

# Contents

# About the Special Issue Editors

**Diego A. Moreno** (Ph.D. in Biological Sciences, University of Granada, Spain, 2000), current position as Scientific Researcher in the Phytochemistry and Healthy Foods Laboratory of the Food Science and Technology Department at CEBAS-CSIC (Spanish National Research Council), member of the Research Group on the Quality, Safety and Bioactivity of Plant Foods (CSIC # 641446). Previous positions as CSIC Staff Scientist (2008–2018), postdoctoral associate in CEBAS-CSIC (2005-2007), and in the Raskin Lab at Rutgers University in New Jersey (USA), from 2001 to 2004. Research interests include foods of plant origin enriched in bioactive and bioavailable phytochemicals and nutrients (glucosinolates, polyphenolics, minerals, vitamins, etc.), designing integrated studies from plants to food and health, and the valorization of agro-food waste to obtain bioactives for industry. Author and coauthor of more than 170 papers in journals (S.C.I.; Scopus-ID 56962713400, h-index 33), Ph.D. adviser, and scientific coordinator and researcher in many competitive projects at the national and international level. Currently active R&D and innovation projects focused on the development of new functional beverages enriched with bioactive phytochemicals to modulate the energy metabolism (inflammation, glucose and lipid metabolism) and improve cognitive performance.

**Nebojsa Ilic** (Ph.D. in Plant biology and biochemistry, University of Maryland, USA, 2000), currently, Senior Research Fellow, Associate Director for Science, Institute for Food Technology, FINS, Novi Sad, Serbia. Performing research in the area of food and health, particularly in the field of functional foods, health supplements and food safety. Discovering and investigating novel functional food ingredients and their incorporation into food products. Chemical, microbiological and sensory analysis of new functional products. Special areas of interest are natural products and bioactive molecules for the treatment of diabetes, cardiovascular disease and inflammation. Participated in several EU funded projects as part of the FP7 and Horizon 2020 programs.

![foods logo] **foods**

**MDPI**

*Editorial*

# Functional and Bioactive Properties of Food: The Challenges Ahead

**Diego A. Moreno [1],* and Nebojsa Ilic [2]**

[1]   CEBAS-CSIC, Phytochemistry and Healthy Foods Lab, Department of Food Science & Technology, Campus Universitario Espinardo–25, E-30100 Espinardo, Murcia, Spain
[2]   Institute of Food Technology, University of Novi Sad, Bul.cara Lazara 1, Novi Sad 21000, Serbia; nebojsa.ilic@fins.uns.ac.rs
*    Correspondence: dmoreno@cebas.csic.es; Tel.: +34-968-396-200 (ext. 6369)

Received: 8 August 2018; Accepted: 24 August 2018; Published: 31 August 2018

With the idea and objective of bringing together new data on biomolecules from fruits, vegetables, wild or medicinal plants, and other organisms (either from land or marine origin) which can exert functional and health-promoting effects through bioactivity beyond the basic nutrient composition, we edited this special issue on "Functional and Bioactive Properties of Food" (URL: http://www.mdpi. com/journal/foods/special_issues/Functional_Bioactive_Properties_Food). The evidence presented in the participating papers is still not enough to demonstrate causality and functionality because the multifactorial conditions of the diseases (which are not related to a single effect or compound) are still a big challenge for this generation of scientists involved in the research of bioactives from foods for nutrition and health.

The principle of "safety first" clearly drives the research on bioactive compounds of natural origin to be incorporated in food and feed formulas for health and wellbeing. Besides bioaccessibility and the demonstration of biological activity, it is clear that an early positive result in any new development would clear up any safety or toxicology issues. One example is the use of ginger rhizome in early pregnancy [1]. More research is needed regarding its efficacy and safety, even though the available data suggests that ginger is a safe and effective treatment for nausea and vomiting in women during early pregnancy [1]. On the other hand, daily food consumption could lead to unexpected exposure of contaminants, such as heavy metals, and the use of plant-derived food formulas could help in the reduction of damage in the human body through different mechanisms of detoxification, as in the use of Selenium-enriched rice grass juice to overcome Cadmium (Cd) contamination in foods [2], even though the evidence is generated in vitro and further developments are expected in the future.

The use of highly-sensitive techniques to fully characterize the components in foods, formulations, and ingredients, as well as to determinate their bioaccessibility, bioavailability, metabolism, and biological activity, is another supporting structure of current and future challenges of the research and development of foods for health. In the separation of compounds in the complex mixture of any given food or formula, as well as in the evaluation of the different effects of growth conditions, agronomical practices, farming, production of any kind, and processing, either at the industrial or the domestic level, must be taken into consideration. The highly-sensitive separation conditions of the fatty acids in digested and processed milk [3], the attention given to the effects of processing on the content of bioactives as the betaine content in gluten-free grains and products [4] has possibilities of fortification, and the necessary phytochemical characterization of bioactive compounds in quite unknown and newly-used fruits of Pitanga [5], are only a few examples of the current trends in the search for bioactive-rich, non-animal-origin foods for health. Nowadays, where everything in terms of foods are connected or related to functionality, a clear, suitable, and effective set of techniques are needed to really highlight the real or potential functionality of a given product, and methods are continuously evolving in terms of accuracy and time consumption, from the preparation of samples to the elucidation of structures and mode of action. Once gathered, the information and data in the discussion should be relevant in terms of the significance of the results, not only at the statistical level but also regarding the physiological and biological relevance of the findings in terms of associative cause–effect between bioactive and bioavailable metabolites and the expected function in target organs. Besides the approaches in improving the scientific work to back-up the results, there is a need and clear evolution in the methodologies too in terms of respect to the environment and hazards in the working environment, with more and more conscious labs using greener alternatives to implement sustainable practices from the farm to the lab and the clinical environments. The design of "smart-foods" also involves nanotechnology to improve the bioavailability, metabolism, and bioactivity of foods, and food components specially designed to target tissues affected from malignancies where the safety and innocuousness of the encapsulated bioactives is the "stone in the road", as seen in the review of Martínez-Ballesta et al. [6] and references therein.

The current trend of producing foods which are aimed at more efficient diets, reduction of intake of products from animal origin, focus on functionality, novel zero-waste processes, and the appreciation of the complex composition and structure of foods and ingredients to be investigated in the context of free-living people using dietary recommendations and dosages compatible with everyday living to treat, managing and modulating the complex conditions of non-communicable diseases from the perspective of prevention is somehow utopic at the present time. However, the aim and will of thousands of research teams from many different areas of science and technology are looking in that direction. Examples of "potential" functional components in foods are included in the special issue of Yamane et al. [7] regarding the probiotics in Kefir, or the healthy lipids in rice Koji extract [8], though the results are obtained through in vitro models—thus being very far from becoming a real functional food or ingredient in the short term. Based on the works related to experiences in humans about functionality, the use of spices in real life did not seem negative [9], but also far from functional. The search for the right delivery system or "attractive" matrix for the consumer, such as cookies enriched with green tea [10], may be acceptable from the organoleptic point of view—however, the biological relevance of the effects on glucose metabolism is scarcely significant. A systematic review on the use of alpha-amylase inhibitors from white bean for body weight and body fat [11] shows only limited reductions, and the limitations of human interventions need better designs and the development of sensible biomarkers to help in the understanding of the connections between foods, nutrition, and health.

In summary, the development of new ingredients and foods enriched with bioactive compounds, having health-promoting characteristics, offers new visions and opportunities for alternative strategies in food production, supply, and consumption. Following this development and the worldwide trend toward foods for nutrition and health, there is a need for a multidimensional and multidisciplinary approach when it comes to connecting foods and their functionality. Among them, of particular importance is performing more in vivo studies showing how functional ingredients and foods actually perform in a living organism, thus demonstrating its bioactivity and functionality. Connected with in vivo studies will be a growing need to perform metabolomic ones that will help to elucidate how these functional ingredients influence metabolism, as well as to explain the possible modes of action. In the end, as most authors stated, further research is required in all and every one of the presented papers in the "Functional and Bioactive Properties of Food", and this assures an exciting time for the researchers in this field and for the general public interested in the relationship between food and health.

**Acknowledgments:** The authors gratefully acknowledge the CYTED Programme for the support of the international collaboration through the Thematic Network AGL 112RT0460 CORNUCOPIA and the Spanish Ministry of Economy and Competitiveness Research Funds through Project AGL2016-75332-C2-1-R (BEBESANO).

**Conflicts of Interest:** The authors declare no conflict of interest. The founding sponsors had no role in the elaboration of the issue or this publication.

# References

1. Stanisiere, J.; MOusset, P.-Y.; Lafay, S. How safe is ginger rhizome for decreasing nausea and vomiting in women during early pregnancy? *Foods* **2018**, *7*, 50. [CrossRef] [PubMed]
2. Chomchan, R.; Siripongvutikorn, S.; Maliyam, P.; Saibandith, B.; Puttarak, P. Protective effect of Selenium-enriched ricegrass juice against Cadmium-Induced toxicity and DNA damage in HEK293 kidney cells. *Foods* **2018**, *7*, 81. [CrossRef] [PubMed]
3. Tunick, M.H.; Van Hekken, D.L. Fatty acid profiles of in vitro digested processed milk. *Foods* **2018**, *6*, 99. [CrossRef] [PubMed]
4. Filipcev, B.; Kojic, J.; Krulj, J.; Bodroza-Solarov, M.; Ilic, N. Betaine in cereal grains and grain-based products. *Foods* **2018**, *7*, 49. [CrossRef] [PubMed]
5. Migues, I.; Baenas, N.; Gironés-Vilaplana, A.; Cesio, M.V.; Heinzen, H.; Moreno, D.A. Phenolic profiling and antioxidant capacity of *Eugenia uniflora* L. (Pitanga) samples collected in different uruguayan locations. *Foods* **2018**, *7*, 67. [CrossRef] [PubMed]
6. Martínez-Ballesta, M.C.; Gil-Izquierdo, A.; García-Viguera, C.; Domínguez-Perles, R. Nanoparticles and controlled delivery for bioactive compounds: Outlining challenges for new "Smart-foods" for health. *Foods* **2018**, *7*, 72. [CrossRef] [PubMed]
7. Yamane, T.; Sakamoto, T.; Nakagaki, T.; Nakano, Y. Lactic acid bacteria from Kefir increase citotoxicity of natural killer cells to tumor cells. *Foods* **2018**, *7*, 48. [CrossRef] [PubMed]
8. Maeda, K.; Ogino, Y.; Nakamura, A.; Nakata, K.; Kitagawa, M.; Ito, S. Identification of rice *Koji* extract components that increase β-glucocerebrosidase levels in human epidermal keratinocytes. *Foods* **2018**, *7*, 94. [CrossRef] [PubMed]
9. Haldar, S.; Lim, J.; Chia, S.C.; Ponnalagu, S.; Henry, C.J. Effects of two doses of curry prepared with mixed spices on postprandial ghrelin and subjective appetite resposnes—A randomized controlled crossover trial. *Foods* **2018**, *7*, 47. [CrossRef] [PubMed]
10. Phongnarisorn, B.; Orfila, C.; Holmes, M.; Marshall, L.J. Enrichment of biscuits with Matcha green tea powder: Its impact on consumer acceptability and acute metabolic response. *Foods* **2018**, *7*, 17. [CrossRef] [PubMed]
11. Udani, J.; Tan, O.; Molina, J. Systematic review and meta-analysis of a proprietary alpha-amylase inhibitor from white bean (*Phaseolus vulgaris* L.) on weight and fat loss in humans. *Foods* **2018**, *7*, 63. [CrossRef] [PubMed]

![foods logo] *foods*                                                      MDPI

*Article*
# Fatty Acid Profiles of In Vitro Digested Processed Milk

Michael H. Tunick [1] and Diane L. Van Hekken [2,*]

[1]  Center for Food and Hospitality Management, Drexel University, 101 North 33rd Street,
     Philadelphia, PA 19104, USA; mht39@drexel.edu
[2]  Dairy and Functional Foods Research Unit, Eastern Regional Research Center, Agricultural Research Service,
     United States Department of Agriculture (USDA), 600 East Mermaid Lane, Wyndmoor, PA 19038, USA
*    Correspondence: Diane.VanHekken@ars.usda.gov; Tel.: +1-215-836-3777

Received: 30 September 2017; Accepted: 3 November 2017; Published: 9 November 2017

**Abstract:** Digestion of milkfat releases some long-chain (18-carbon) fatty acids (FAs) that can provide health benefits to the consumer, yet because they are found in small amounts and can be difficult to identify, there is limited information on the effects that common fluid milk processing may have on the digestibility of these FAs. This study provides FA profiles for raw and combinations of homogenized and/or heat-treated (high and ultra-high temperature pasteurization) milk, before and after in vitro digestion, in order to determine the effects of processing on the digestibility of these healthy fatty acids. Use of a highly sensitive separation column resulted in improved FA profiles that showed that, when milk was subjected to both pasteurization and homogenization, the release of the 18-carbon FAs, oleic acid, linoleic acid (an omega-6 FA), rumenic acid (a conjugated linoleic acid, CLA), and linolenic acid (an omega-3 FA) tended to be higher than with either pasteurization or homogenization, or with no treatment. Milk is noted for containing the omega-3 FAs and CLAs, which are associated with positive health benefits. Determining how processing factors may impact the components in milk will aid in understanding the release of healthy FAs when milk and dairy foods are consumed.

**Keywords:** milk; fatty acids; digestion; pasteurization; homogenization; omega-3 fatty acids; conjugated linoleic acid

---

## 1. Introduction

Milk is an important part of the diet of many people. Digestion of milk fat, 98% in the form of triacylglycerides (TAG), releases bioactive fatty acids that are absorbed by the small intestine providing a multitude of health benefits. The 18-carbon polyunsaturated fatty acids (PUFAs), making up about 2.3% of the FAs in milk, have been linked to improving mental health, reducing inflammation, inhibiting some cancers, and preventing many chronic illnesses such as cardiovascular disease, diabetes, and obesity [1–5]. The major 'healthy' FAs are the omega-3 FAs and conjugated linoleic acids (CLA). The primary omega-3 FA in milk is α-linolenic acid (*cis*-9, *cis*-12, *cis*-15 18:3). The CLAs are isomers of linoleic acid (*cis*-9, *cis*-12 18:2) with rumenic acid (*cis*-9, *trans*-11 18:2), one of the most bioactive of the CLAs, comprising 92% of the total CLAs in milk [3].

Overall, milk is considered to be a highly digestible food. Yet there are many processing factors, such as homogenization and high heat treatments, which are known to alter the stability and structure of components in milk [6] that may influence digestibility [7]. Homogenization at pressures of 10–25 MPa ruptures the protective milk fat globular membrane (MFGM) from the fat droplet and reduces it to smaller droplets (1–3 μm) [8]. High heat treatments are used to eliminate pathogens and reduce spoilage bacteria in order to extend the shelf life of milk. Common methods include high temperature, short time (HTST) pasteurization at 72 °C for 15 s or ultra-high temperature (UHT) pasteurization at 135 °C for 2 s.

Although attempts to study the effects of processing on milkfat digestion have been made, it is difficult to compare results because of the variety of methods employed. One in vivo study used fasting rats fed raw, pasteurized, or pasteurized/homogenized cream. Results showed that pasteurized cream under gastric conditions had the lowest levels of long-chain (≥14-carbons) FAs. Pasteurized/homogenized cream under intestinal conditions had the lowest levels of medium- and long-chain (≥8-carbons) FAs [9]. Another approach used human gastric juices containing pepsin and lipases as well as porcine pancreatin and bile salts in the in vitro digestion of milk samples with different-sized lipid droplets. Results showed that the small native lipid droplets released more FAs under gastric and intestinal conditions than larger native or smaller homogenized droplets. The size of the droplet also influenced the FAs released, which suggested that the size of the droplet may alter the positioning of the FAs at the droplet surface (enzyme access interface) [10].

Previous research in our laboratory used yet another approach based on the simulated fasting human gastro-intestinal model, which uses porcine pepsin and pancreatin and bovine bile extract [11] to study the digestibility of raw and processed milk [12]. Although free fatty acids (FFAs) were not released during gastric digestion, raw and homogenized samples released more FFAs than the pasteurized samples during intestinal digestion. However, the rate of FA release began to slow 30 min after lipase addition, and flattened after 90 min because of product inhibition. Comparison of FA profiles prior to digestion and after 3 h of gastro-intestinal digestion showed that stearic acid (18:0) and oleic acid (18:1) were released faster than the other FAs. All processing treatments showed similar trends in digestion for the same FA, but linoleic acid (*cis*-9, *cis*-12 18:2), its CLA isomers, and α-linolenic acid (*cis*-9, *cis*-12, *cis*-15 18:3) were not reported. Similar digestion trends were found in a follow up study that compared the quantity of FFAs released during in vitro gastro-intestinal digestion of raw and processed milk from grazing organic and confined conventional cows [13]. In that study, over half of the FFAs released in the intestinal phase were generated in the first 15 min after lipase addition. Milk from confined conventional herds, which had higher levels of palmitic (16:0) and stearic (18:0) acids, released more FFAs compared to the grazing organic herd. These FAs are typically located on the sn-1 and sn-3 TAG positions that are preferred by pancreatic lipase.

The diversity of the approaches used to study the digestion of milk offers unique opportunities to understand different aspects of the digestion process. It is imperative that the FA profiles include the healthy C18 FAs, not only to provide baselines for future research, but to aid in understanding the release of bioactive FAs in dairy foods. This work, stemming from our organic milk study [13], (1) provides complete and accurate FA profiles for milk and digested samples using a more sensitive column and (2) uses the FA profiles to evaluate the effect of homogenization and/or heat pasteurization on the gastro-intestinal digestion of milkfat.

## 2. Materials and Methods

### 2.1. Processing of Milk

Three shipments of raw milk (38 L) were obtained from an actively grazing certified organic herd located in Berks County, PA between mid-July and early September and processed that afternoon as described by Van Hekken et al. [13] in our dairy pilot plant facilities. Briefly, milk was standardized to 3.25% fat and divided into five portions. One portion remained unheated raw (R) milk control. Another portion was high temperature, short time (HTST) pasteurized (P) at 72 °C for 15 s using a Universal Pilot Plant (UPP, Waukesha Cherry-Burrell, Philadelphia, PA, USA). A third portion was HTST pasteurized and homogenized (HP) using the two-stage homogenizer (17.2/10.3 MPa) attached to the UPP. Another portion of the raw milk was warmed to 60 °C to liquefy the fat before being homogenized (H) in the off-line UPP homogenizer. The last portion was ultra-high temperature pasteurized at 135 °C for 2 s using a FT 74 P/T HTST/UHT Processing System (Armfield, Denison, IA, USA) and homogenized (HU) in the off-line UPP homogenizer. Milk samples were stored at 4 °C until digested and aliquots of milk for FA analysis were frozen at −35 °C until analyzed.

## 2.2. Digestion of Milk

Digestion of milk was based on the in vitro human fasting model by Gallier et al. [14] as modified by Tunick et al. [12] and described by Van Hekken et al. [13]. Raw milk samples were digested the next day and all treatments were digested within 3 days after processing; digestions were conducted once. Briefly, 50 mL of warmed (37 °C) simulated gastric fluid [11] consisting of 0.2% NaCl and 0.32% pepsin (P7000 from porcine gastric mucosa, Sigma-Aldrich, St. Louis, MO, USA), pH 1.2, was added to 100 mL of warmed (37 °C) milk, and adjusted to pH 1.5 with 6 M HCl (Fisher Scientific, Fair Lawn, NJ, USA). The mixture was constantly shaken in a water bath at 37 °C for 60 min to produce a gastric digest. A mixture of 1:1 gastric digest: simulated intestinal fluid (SIF) [11] containing 0.68% $KPO_4$ (Fisher) was adjusted to pH 7.0 with 0.2 N NaOH (Fisher). Bile salts (B8631 porcine bile extract, Sigma-Aldrich, St. Louis, MO, USA) were added to a concentration of 5 mg/mL SIF, along with porcine pancreatin (P1750, Sigma-Aldrich) to a concentration of 1.6 mg/mL SIF. The pH was not adjusted during digestion. Aliquots of the intestinal digests were taken after 120 min and centrifuged at $5000 \times g$ for 30 min at 10 °C (Avanti J-301; Beckman Coulter, Inc., Brea, CA, USA) to obtain the lipid portion, which was immediately frozen at −35 °C (REVCO TM DxF; Thermo Scientific, Inc., Waltham, MA, USA) until analyzed.

## 2.3. Fatty Acid Profiles

Profiles of the FAs still bound to the TAG in each processed milk and digested samples were obtained based on a modified version of the procedure described by Tunick et al. [12]. Briefly, the FAs were converted to methyl esters (FAME) by reacting with sodium methoxide (Sigma-Aldrich) (free FAs were not detected by this method) and 0.5 µL samples were injected into a Trace 1300 gas chromatograph equipped with a flame ionization detector set at 270 °C (Thermo Scientific, Waltham, MA, USA) and a CP-Sil 88 column (100 m × 0.25 mm; Agilent Technologies, Wilmington, DE, USA). Helium flow was 1.0 mL/min and the injector was 20 mL/min split flow with the inlet temperature at 250 °C. The initial oven temperature was 80 °C and was increased at 4 °C/min to 220 °C and held for 5 min; the oven temperature was then increased at 4 °C/min to 240 °C and held for the remaining 10 min: The instrument's software was used to calculate percentages of FA. The internal standard added before esterification was glyceryl trinonanoate (Sigma-Aldrich), and reference standards consisted of 4–24-carbon methyl esters and conjugated methyl linoleate (Nu-Chek-Prep, Elysian, MN, USA). Data collection started with caprylic acid (8:0).

## 2.4. Statistical Analysis

FA profile means for each shipment were used to determine the effect of milk processing, digestion, and the processing-digestion interaction on the FAs present in the samples. Statistical software was used to conduct analysis of variance (ANOVA) using the PROC MIXED statements and the Bonferroni LSD test to determine significant differences ($p < 0.05$) (Statistical Analysis System (SAS) for Windows, version 9.4; SAS Institute Inc., Cary, NC, USA). The repeatability of the digestion method was determined using the PROC GLM (SAS) to obtain the coefficient of variance (CV) and standard deviation (SD) for the undigested ($n = 15$) and digested milk ($n = 11$) samples.

## 3. Results and Discussion

### 3.1. GC System

Use of the longer (100 m) fused silica column coated with the highly polar cyanoalkyl polysiloxane stationary phase, CP Sil 88, resulted in finer resolution of peaks and identification of more monounsaturated FAs (MUFAs) and odd-numbered FAs. In total, 8 saturated FAs (SFA), 6 MUFAs, and 3 of the healthy PUFAs (Table 1) were quantified, which is an improvement over our previous GC system that used a 60 m fused silica column with a stationary phase of biscyanopropyl/cyanopropylphenyl siloxane [12]. The major CLA in milk, rumenic acid (*cis*-9, *trans*-11 18:2), was quantified, whereas the other CLA isomers were not detected. Although 400 different FAs that have been identified in milk, only 20 FAs make up 90% of the FAs found in milk [15]. The FAs present in trace amounts, such as the other CLAs, will be difficult to isolate and identify. Further modification of the FA analysis protocol, both in sample preparation and the GC method program, may resolve overlapping peaks to identify even more FAs, including other CLAs and omega-3 FAs.

**Table 1.** Fatty acid concentrations (g/100 g) for differently processed milk before and after a 3 h gastro-intestinal digestion.

| Fatty Acid | Common Name | R | H | P | HP | HU | SE |
|---|---|---|---|---|---|---|---|
| *Undigested milk* | | | | | | | |
| 8:0 | Caprylic acid | 1.20 [a] | 1.38 [a] | 1.17 [a] | 1.58 [a] | 1.56 [a] | 0.201 |
| 10:0 | Capric acid | 2.95 [abc] | 3.57 [ab] | 3.07 [abc] | 2.63 [abc] | 3.72 [a] | 0.289 |
| 12:0 | Lauric acid | 3.39 [a] | 4.15 [a] | 3.67 [a] | 3.67 [a] | 4.32 [a] | 0.274 |
| 14:0 | Myristic acid | 11.25 [c] | 13.35 [c] | 11.89 [c] | 12.28 [c] | 12.57 [c] | 0.477 |
| 14:1 | Myristoleic acid | 1.22 [a] | 1.52 [a] | 1.13 [a] | 1.30 [a] | 1.07 [a] | 0.352 |
| 15:0 | | 1.19 [a] | 1.42 [a] | 1.32 [a] | 1.31 [a] | 1.28 [a] | 0.284 |
| 15:1 | | 0.19 [c] | 0.21 [bc] | 0.22 [bc] | 0.22 [bc] | 0.24 [bc] | 0.030 |
| 16:0 | Palmitic acid | 29.10 [b] | 30.57 [b] | 29.69 [b] | 28.52 [b] | 28.46 [b] | 1.038 |
| 16:1 | Palmitoleic acid | 1.30 [a] | 1.53 [a] | 2.31 [a] | 1.79 [a] | 2.69 [a] | 0.312 |
| 17:0 | Margaric acid | 0.72 [a] | 0.69 [a] | 0.79 [a] | 0.74 [a] | 0.74 [a] | 0.128 |
| 17:1 | | 0.30 [a] | 0.31 [a] | 0.30 [a] | 0.30 [a] | 0.29 [a] | 0.180 |
| 18:0 | Stearic acid | 13.85 [a] | 11.27 [a] | 11.94 [a] | 12.68 [a] | 10.59 [a] | 1.292 |
| *cis*-9 18:1 | Oleic acid | 23.89 [a] | 21.14 [a] | 23.54 [a] | 24.15 [a] | 24.23 [a] | 1.034 |
| *trans*-11 18:1 | Vaccenic acid | 3.85 [a] | 3.39 [ab] | 3.50 [ab] | 3.65 [ab] | 3.47 [ab] | 0.172 |
| *cis*-9, *cis*-12 18:2 | Linoleic acid | 3.74 [a] | 3.72 [a] | 3.64 [a] | 3.52 [a] | 3.36 [a] | 0.203 |
| *cis*-9, *trans*-11 18:2 | Rumenic acid | 0.97 [a] | 0.97 [a] | 0.93 [ab] | 0.92 [ab] | 0.81 [ab] | 0.056 |
| *cis*-9, *cis*-12, *cis*-15 18:3 | α-Linolenic acid | 0.77 [a] | 0.78 [a] | 0.67 [ab] | 0.69 [ab] | 0.63 [abc] | 0.044 |
| *Digested milk* | | | | | | | |
| 8:0 | Caprylic acid | 1.47 [a] | 0.92 [a] | 0.98 [a] | 0.55 [a] | 0.76 [a] | 0.201 |
| 10:0 | Capric acid | 2.16 [abc] | 3.79 [ab] | 2.16 [a] | 1.94 [bc] | 1.50 [c] | 0.289 |
| 12:0 | Lauric acid | 4.24 [a] | 4.64 [a] | 3.67 [abc] | 4.11 [a] | 3.75 [a] | 0.274 |
| 14:0 | Myristic acid | 18.13 [ab] | 17.50 [b] | 17.48 [a] | 20.67 [a] | 19.62 [ab] | 0.477 |
| 14:1 | Myristoleic acid | 2.20 [a] | 2.58 [a] | 2.31 [b] | 2.00 [a] | 2.15 [a] | 0.352 |
| 15:0 | | 1.40 [a] | 1.88 [a] | 2.09 [a] | 1.99 [a] | 2.12 [a] | 0.284 |
| 15:1 | | 0.22 [bc] | 0.22 [bc] | 0.49 [a] | 0.37 [ab] | 0.31 [abc] | 0.030 |
| 16:0 | Palmitic acid | 37.60 [a] | 38.50 [a] | 39.82 [a] | 41.59 [a] | 43.77 [a] | 1.038 |
| 16:1 | Palmitoleic acid | 2.58 [a] | 2.05 [a] | 2.33 [a] | 2.21 [a] | 1.74 [a] | 0.312 |
| 17:0 | Margaric acid | 1.11 [a] | 1.22 [a] | 0.88 [a] | 0.94 [a] | 1.00 [a] | 0.128 |
| 17:1 | | 0.81 [a] | 0.33 [a] | 0.45 [a] | 0.75 [a] | 0.81 [a] | 0.180 |
| 18:0 | Stearic acid | 7.59 [a] | 6.97 [a] | 8.06 [a] | 7.76 [a] | 8.19 [a] | 1.292 |
| *cis*-9 18:1 | Oleic acid | 13.21 [b] | 13.47 [b] | 12.54 [b] | 9.70 [b] | 8.82 [b] | 1.034 |
| *trans*-11 18:1 | Vaccenic acid | 3.36 [abc] | 2.22 [c] | 2.86 [abc] | 2.82 [bc] | 2.90 [abc] | 0.172 |
| *cis*-9, *cis*-12 18:2 | Linoleic acid | 2.89 [ab] | 2.76 [ab] | 2.81 [ab] | 1.99 [b] | 1.97 [b] | 0.203 |
| *cis*-9, *trans*-11 18:2 | Rumenic acid | 0.60 [bc] | 0.59 [bc] | 0.57 [bc] | 0.33 [c] | 0.33 [c] | 0.056 |
| *cis*-9, *cis*-12, *cis*-15 18:3 | α-Linolenic acid | 0.42 [bcd] | 0.38 [cd] | 0.41 [bcd] | 0.28 [d] | 0.27 [d] | 0.044 |

Processing treatments included: raw, R; homogenized, H; pasteurized, P; homogenized and pasteurized, PU; and homogenized and UHT pasteurized, HU. [a–d] Means for the same fatty acid not sharing the same letter are significantly different ($p < 0.05$). Standard error, SE.

## 3.2. FA Profiles

Table 1 shows the types and levels of FAs in the milk and digested samples. The FA profiles of the milk prior to digestion were similar to those of previously published studies of milk from actively grazing cows in the USA [16–18]. All of the profiles were dominated by palmitic (16:0) and oleic (18:1) acids, comprising over 50% of the FAs. The myristic (14:0) and stearic (18:0) acids comprised about 20% of the FAs.

Rumenic acid (*cis*-9, *trans*-11 18:2) was present at nearly 1%, and none of the other 20 isomers reported in the milk [19] were detected. Statistical analysis of the milk and digested milk data resulted in coefficient of variance of 10.9 ± 0.44 and 13.6 ± 0.52, respectively, for the overall repeatability of the experiment.

The digestion system we used did not add lipase until the intestinal phase [14], and we did not detected any FFAs in the digest until then. Other studies that used the in vivo rat model [9] or human gastric juices in an in vitro model [10] reported the release of FFAs during the gastric phase. It has been suggested that gastric lipolysis enhances the efficiency of pancreatic lipase [4,10].

The profiles changed markedly with simulated intestinal digestion, due to the lipolytic activity of the added pancreatic lipases supported by the bile salts. The concentrations of many of the 18-carbon FAs decreased significantly ($p < 0.05$), which indicated the release of FFAs into the digest (Table 1). The amounts of myristic (14:0) and palmitic (16:0) acids in the digested samples appeared to increase (Table 1), although they were being released but at a slower rate than the other FAs. The slower release rate resulted in their higher concentration in the recovered digested fat sample. In bovine milk, 45–65% of myristic (14:0) and palmitic (16:0) acids are located in the sn-2 (center) position of the triacylglyceride molecule [20], partially protecting it from lipolysis. In contrast, the 18-carbon FAs (18:0, 18:1, 18:2 and 18:3) are situated in the sn-1 and sn-3 (outer) positions [20], allowing them to be readily released.

The reductions in 18-carbon FAs may be more easily visualized by noting their ratios to myristic acid before and after digestion (Table 2). The ratios for 18:1, 18:2 and 18:3 substantially decreased ($p < 0.05$) with digestion. Homogenized heat treated (HP and HU) samples tended to have lower ratios for oleic (18:1), linoleic (*cis*-9, *cis*-12 18:2), rumenic (*cis*-9, *cis*-11 18:2), and α-linolenic (*cis*-9, *cis*-12, *cis*-15 18:3) acids. These levels were reduced by more than half when the HP and HU samples were digested, and their ratios to myristic acid (14:0) were significantly lower than the corresponding ratios of the raw, homogenized and pasteurized samples. Homogenization or HTST pasteurization alone did not produce these effects, suggesting that the combination of two treatments may increase digestibility of these four FAs and will require further bench and clinical research to elucidate.

**Table 2.** Ratios of 18-carbon fatty acids to myristic acid (14:0) for differently processed milk before and after a 3 h gastro-intestinal digestion.

| Fatty Acid | Common Name | R | H | P | HP | HU | SE |
|---|---|---|---|---|---|---|---|
| Undigested milk | | | | | | | |
| 18:0 | Stearic acid | 1.23 [a] | 0.85 [ab] | 1.02 [ab] | 1.03 [ab] | 0.85 [ab] | 0.116 |
| *trans*-11 18:1 | Vaccenic acid | 0.34 [a] | 0.26 [abc] | 0.30 [a] | 0.30 [a] | 0.28 [ab] | 0.015 |
| *cis*-9 18:1 | Oleic acid | 2.12 [a] | 1.59 [a] | 1.98 [a] | 1.98 [a] | 1.93 [a] | 0.104 |
| *cis*-9, *cis*-12 18:2 | Linoleic acid | 0.33 [a] | 0.28 [ab] | 0.31 [a] | 0.29 [a] | 0.27 [abc] | 0.018 |
| *cis*-9, *trans*-11 18:2 | Rumenic acid | 0.086 [a] | 0.073 [a] | 0.078 [a] | 0.075 [a] | 0.065[ab] | 0.005 |
| *cis*-9, *cis*-12, *cis*-15 18:3 | α-Linolenic acid | 0.069 [a] | 0.058 [a] | 0.056 [a] | 0.056 [a] | 0.050 [a] | 0.003 |
| Digested milk | | | | | | | |
| 18:0 | Stearic acid | 0.42 [b] | 0.40 [b] | 0.47 [ab] | 0.38 [b] | 0.42 [b] | 0.116 |
| *trans*-11 18:1 | Vaccenic acid | 0.19 [bcd] | 0.13 [d] | 0.16 [cd] | 0.14 [d] | 0.15 [d] | 0.015 |
| *cis*-9 18:1 | Oleic acid | 0.73 [b] | 0.77 [b] | 0.72 [b] | 0.47 [b] | 0.44 [b] | 0.104 |
| *cis*-9, *cis*-12 18:2 | Linoleic acid | 0.16 [bcd] | 0.16 [cd] | 0.16 [bcd] | 0.10 [d] | 0.10 [d] | 0.018 |
| *cis*-9, *trans*-11 18:2 | Rumenic acid | 0.034[bc] | 0.035[bc] | 0.034[bc] | 0.016 [c] | 0.017 [c] | 0.005 |
| *cis*-9, *cis*-12, *cis*-15 18:3 | α-Linolenic acid | 0.024 [b] | 0.021 [b] | 0.024 [b] | 0.013 [b] | 0.014 [b] | 0.003 |

Processing treatments included: raw, R; homogenized, H; pasteurized, P; homogenized and pasteurized, PU; and homogenized and UHT pasteurized, HU. [a–d] Means for the same fatty acid not sharing the same letter are significantly different ($p < 0.05$). Standard error, SE.

Overall, our results are similar to findings from other digestion studies [9,12]. The in vivo rat study also reported that pasteurized/homogenized cream released the most FAs containing ten or more carbons, but only reported the 18-carbon FAs stearic (18:0), oleic (18:1), and linoleic (*cis*-9, *cis*-12 18:2) acids [9]. An earlier in vitro study from our lab reported greater release of stearic (18:0) and oleic (18:1) acids from homogenized and heat processed milk [12], but did not report results for the PUFAs.

Unlike our study, in vitro studies using human enzymes showed higher levels of digestion from smaller fat droplets, but found homogenization and UHT heat treatment did not enhance milk digestion [10]. They reported that the size of the lipid droplet and its surface (covered with the MFGM or casein/whey proteins) were key to lipid digestion. They also suggested that pancreatic lipases had a lower affinity to the casein coating that forms around the smaller homogenized fat droplets, thereby limiting the release of FAs.

## 4. Conclusions

Findings from this study showed that the use of the longer, highly polar GC column improved the resolution, identification, and quantification of the FAs present in milk and the in vitro digested milk samples. Release of the FAs was easily tracked by comparing the FA profiles before and after digestion. The 18-carbon FAs were rapidly released during intestinal digestion, although care must be taken when interpreting all FA release results to take into account the distribution of the FAs in the diminishing quantity of fat at the end of digestion. While our results suggest that pasteurizing and homogenizing milk may lead to better digestibility of healthy FAs, more research is needed using animal and human models to further investigate this phenomenon.

The diversity of the systems that have been used to digest milk offers unique opportunities to understand certain aspects of the digestion process, which, when combined, will improve our understanding of the effects that processing may have on the human health benefits from consuming milk-based foods.

**Acknowledgments:** The authors thank Daxi Ren, Zhejiang University, China for carrying out the digestion experiments and Anthony Bucci, Agricultural Research Service, Dairy & Functional Foods Research Unit for his contribution in preparing this manuscript and Bryan Vinyard, Agricultural Research Service, Northeast Area Statistics Group, for his statistical advise. Mention of commercial products in this publication is solely for the purpose of providing specific information and does not imply recommendation or endorsement by the USDA. USDA is an equal opportunity provider and employer.

**Author Contributions:** M.H. Tunick contributed to the experimental design and generation of samples, prepared the fatty acid methyl esters, conducted GC-FID analysis of lipid samples, and prepared the initial manuscript. D.L. Van Hekken contributed to the experimental design and generation of the processed milk samples, extracted lipids from the frozen samples, conducted the statistical analysis of data, and prepared the final version of the manuscript for submission.

**Conflicts of Interest:** The authors declare no conflict of interest.

## References

1. El-Badry, A.M.; Graf, R.; Clavien, P.A. Omega 3–Omega 6: What is right for the liver? *J. Hepatol.* **2007**, *47*, 718–725. [CrossRef] [PubMed]
2. Riediger, N.D.; Othman, R.A.; Suh, M.; Moghadasian, M.H. A systemic review of the roles of n-3 fatty acids in health and disease. *J. Am. Diet. Assoc.* **2009**, *109*, 668–679. [CrossRef] [PubMed]
3. Koba, K.; Yanagita, T. Health benefits of conjugated linoleic acid (CLA). *Obs. Res. Clin. Pract.* **2014**, *8*, e525–e532. [CrossRef] [PubMed]
4. Pereira, P.C. Milk nutritional composition and its role in human health. *Nutrition* **2014**, *30*, 619–627. [CrossRef] [PubMed]
5. Simopoulos, A.P. An increase in the omega-6/omega-3 fatty acid ratio increases the risk for obesity. *Nutrients* **2016**, *8*, 128. [CrossRef] [PubMed]
6. Michalski, M.C.; Januel, C. Does homogenization affect the human health properties of cow's milk? *Trends Food Sci. Technol.* **2006**, *17*, 423–437. [CrossRef]
7. Singh, H.; Ye, A.; Horne, D. Structuring food emulsions in the gastrointestinal tract to modify lipid digestion. *Progress Lipid Res.* **2009**, *48*, 92–100. [CrossRef] [PubMed]
8. Tetra Pak. *Dairy Processing Handbook*, 3rd ed.; Tetra Pak Processing Systems AB: Lund, Sweden, 2015; pp. 131–138. ISBN 87-104227-232.
9. Gallier, S.; Zhu, X.Q.; Rutherfurd, S.M.; Ye, A.; Moughan, P.J.; Singh, H. In vivo digestion of bovine milk fat globules: Effect of processing and interfacial structural changes. II. Upper digestive tract digestion. *Food Chem.* **2013**, *141*, 3215–3223. [CrossRef] [PubMed]

10. Garcia, C.; Antona, C.; Robert, B.; Lopez, C.; Armand, M. The size and interfacial composition of milk fat globules are key factors controlling triglycerides bioavailability in simulated human gastro-duodenal digestion. *Food Hydrocoll.* **2014**, *35*, 494–504. [CrossRef]
11. US Pharmacopeia. *The National Formulary 30*; US Pharmacopeia Board of Trustees: Rockville, MD, USA, 2012; Volume 1, p. 5778.
12. Tunick, M.H.; Ren, D.X.; Van Hekken, D.L.; Bonnaillie, L.; Paul, M.; Kwoczak, R.; Tomasula, P.M. Effect of heat and homogenization on in vitro digestion of milk. *J. Dairy Sci.* **2016**, *99*, 4124–4139. [CrossRef] [PubMed]
13. Van Hekken, D.L.; Tunick, M.H.; Ren, D.X.; Tomasula, P.M. Comparing the effect of homogenization and heat processing of the properties and in vitro digestion of milk from organic and conventional dairy herds. *J. Dairy Sci.* **2017**, *100*, 6042–6052. [CrossRef] [PubMed]
14. Gallier, S.; Singh, H. Behavior of almond oil bodies during in vitro gastric and intestinal digestion. *Food Funct.* **2012**, *3*, 547–554. [CrossRef] [PubMed]
15. Masson, H.L. Fatty acids in bovine milk fat. *Food Nutr. Res.* **2008**, *52*, 1821–1823. [CrossRef]
16. Lock, A.L.; Bauman, D.E. Modifying Milk Fat Composition of Dairy Cows to Enhance Fatty Acids beneficial to human health. *Lipids* **2004**, *39*, 1197–1206. [CrossRef] [PubMed]
17. Benbrook, C.M.; Butler, G.; Latif, M.A.; Leifert, C.; Davis, D.R. Organic production enhances milk nutritional quality by shifting fatty acid composition: A United States-wide, 18-month study. *PLoS ONE* **2013**, *9*, e82429–e82435. [CrossRef] [PubMed]
18. Tunick, M.H.; Paul, M.; Ingham, E.R.; Karreman, H.J.; Van Hekken, D.L. Differences in milk characteristics between a cow herd transitioning to organic versus milk from a conventional dairy herd. *Int. J. Dairy Technol.* **2015**, *68*, 511–518. [CrossRef]
19. Kraft, J.; Collomb, M.; Möckel, P.; Sieber, R.; Jahreis, G. Differences in CLA isomer distribution of cow's milk lipids. *Lipids* **2003**, *38*, 657–664. [CrossRef] [PubMed]
20. Jensen, R.G. The composition of bovine milk lipids: January 1995 to December 2000. *J. Dairy Sci.* **2002**, *85*, 295–350. [CrossRef]

*foods*

MDPI

*Article*

# Enrichment of Biscuits with Matcha Green Tea Powder: Its Impact on Consumer Acceptability and Acute Metabolic Response

**Benjapor Phongnarisorn [1,2], Caroline Orfila [1,*], Melvin Holmes [1] and Lisa J. Marshall [1,*]**

[1] School of Food Science and Nutrition, University of Leeds, Leeds LS2 9JT, UK; nbjp459@hotmail.com (B.P.); prcmjh@leeds.ac.uk (M.H.)

[2] Faculty of Agricultural Technology, Phuket Rajabhat University, Phuket 83000, Thailand

* Correspondence: c.orfila@leeds.ac.uk (C.O.); L.J.Marshall@leeds.ac.uk (L.J.M.); Tel.: +44-113-343-2966

Received: 5 January 2018; Accepted: 26 January 2018; Published: 1 February 2018

**Abstract:** Matcha green tea powder (MGTP) is made with finely ground green tea leaves that are rich in phytochemicals, most particularly catechins. Shortbread biscuits were enriched with MGTP and evaluated for consumer acceptability and potential functional health properties. Baking decreased the content of total catechins by 19% compared to dough, although epimerization increased the amount of (+)-gallocatechin gallate at the expense of other catechins such as (−)-epigallocatechin gallate. Consumer acceptability tests using a 9-point hedonic scale showed that consumers preferred enriched biscuits with low content of MGTP (2 g of MGTP 100 $g^{-1}$ of flour), and an increase of sugar content did not significantly improve the acceptability of MGTP-enriched biscuits. Overall, enrichment of biscuits with MGTP did not significantly affect the postprandial glucose or triglyceride response (area under curve) compared to non-enriched biscuits consumed with water or MGTP drink. Enriching biscuits with Matcha green tea is acceptable to consumers, but may not bring significant postprandial effects.

**Keywords:** Matcha green tea powder; catechin stability; consumer acceptability; acute metabolic response; functional food

---

## 1. Introduction

Catechins are the main polyphenols found in green tea. As shown in Figure 1, catechins have a common carbon backbone with variations in the substitutes at the C-3 and C-5′ positions [1]. The four major catechins found in green tea are (−)-epigallocatechin (EGC), (−)-epicatechin (EC), (−)-epigallocatechin gallate (EGCG), and (−)-epicatechin gallate (ECG). Their trans epimer forms, namely (+)-catechin, (+)-gallocatechin gallate (GCG), (+)-gallocatechin (GC), and (+)-catechin gallate (CG), are found in minor amounts. The epimerization from cis to trans is reversible and can occur when cis catechins are exposed to high temperatures [2]. EGCG is the most abundant catechin in green tea leaves and contributes to 50% of total catechins in tea leaves. Hence, it is been used as a quality indicator in green tea products [3]. It is also found at low levels in a range of foods including apples, red berries, nuts, and legumes.

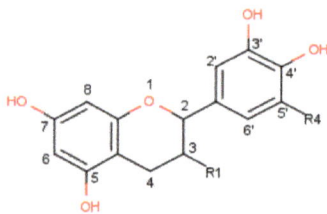

| | R₁ | R₄ |
|---|---|---|
| Catechin | OH | H |
| Epicatechin (EC) | OH | H |
| Epigallocatechin (EGC) | OH | OH |
| Epicatechin gallate (ECG) | Gallate | H |
| Epigallocatechin gallate (EGCG) | Gallate | OH |

**Figure 1.** Structure of the main catechins found in green tea.

Consumer research has suggested that even though green tea is unfamiliar to most non-Asian consumers, their desire to drink green tea is enhanced due to its perceived health benefits [4,5]. Moreover, there are also the demands for convenience products (i.e., biscuits) with value-added components [6,7]. Incorporating green tea powder into bakery products may mask the bitterness or astringency of green tea perceived by consumers. A few studies have incorporated green tea into bakery products and have shown effects on the physicochemical, colour, textural, and shelf life properties [8–10]. Studies found good sensory acceptability of cakes enriched with up to 20% green tea substituting wheat flour [9]. To date, there have been no reports on the potential physiological effects of green tea-enriched food products.

The catechins found in green tea have been shown to have potential roles in preventing cancer development, diabetes, and cardiovascular diseases [11,12]. Zhong et al. [13] found a 25% decrease in carbohydrate absorption after a consumption of green tea extract containing 300 mg of EGCG and 100 mg of ECG with a carbohydrate meal, suggesting that green tea may reduce the amount of glucose absorbed into the bloodstream, which in turn may lower the risk of developing diabetes [12,13]. Moreover, a review by Koo and Noh [14] suggested that green tea reduces the absorption of dietary fat by interfering with the processes of lipid digestion in the intestine. Studies using animal models have shown that green tea compounds can slow down lipid absorption by inhibiting the activity of pancreatic lipase [15,16]. Unno et al. [16] showed that 224 mg of green tea catechins consumed as tea infusion after meal consumption can lower the concentration of lipids in the blood by 15% and therefore may lower the risk of developing cardiovascular diseases. Studies so far have used green tea in the form of pure extracts or tea infusions to study the potential health effects, but none have used it as part of a functional food.

The aims of this study were to (1) investigate the effect of baking on the stability of catechins; (2) evaluate the sensorial acceptability of shortbread biscuits enriched with Matcha green tea powder (MGTP) and assess the effect of sugar addition on acceptability; and (3) investigate the effect of MGTP on the postprandial glucose and triglyceride response in human volunteers. We discuss the potential of MGTP-enriched biscuits as functional foods.

## 2. Materials and Methods

### 2.1. Materials

Matcha factory® Matcha green tea powder (13.2% of total phenolic content, 10% total catechin) was purchased from Chah Ltd. (Solihull, UK). Tesco® plain flour, Tate & Lyle® caster sugar, and Anchor® unsalted butter were purchased from Tesco supermarket (Leeds, UK).

HPLC (High Performance Liquid Chromatography)-grade acetonitrile and ethanol were purchased from Fisher Scientific Co. Ltd. (Leicestershire, UK). HPLC-grade methanol, (−)-epicatechin (EC), and caffeine were purchased from Sigma Aldrich Co. Ltd. (St. Louis, MI, USA). (−)-Epicatechin gallate (ECG), (−)-epigallocatechin (EGC), and (−)-epigallocatechin gallate (EGCG) were obtained from Cambridge Bioscience Ltd. (Cambridge, UK). (+)-Gallocatechin gallate (GCG) was purchased from Insight

Biotechnology Co. Ltd. (Middlesex, UK). All water refers to deionized Millipore water, Millipore Ltd. (Hertfordshire, UK).

## 2.2. Preparation of Shortbread Incorporated with MGTP

The dough was prepared by beating butter (83.3 g) and sugar (at three different levels, 25, 30, or 35 g 100 g$^{-1}$ of flour) together with a mixer (KM336 Chef Classic, Kenwood Ltd., Havant, UK) for 5 min at speed 4 (180 rpm), and then flour (100 g) and MGTP were added (at the level of 2, 4, or 6 g 100 g$^{-1}$ of flour), followed by mixing for 5 min at minimum speed (50 rpm). The dough was wrapped in clingfilm and rested for 1 h at 4 °C. The dough was rolled to 0.4 mm thickness, and then the biscuits were cut into a circle shape (4.8 cm diameter) and rested for 10 min before being placed in a conventional oven (BC190.2TCSS, Baumatic Ltd., Merthyr Tydfil, UK) at 180 °C for 15 min. After baking, the biscuits were left to cool down at room temperature.

The weight of biscuits was measured before and after baking to calculate weight loss during baking. Samples of each biscuit were prepared for sensory evaluation. For HPLC analysis, the dough and shortbread biscuit samples (made with 25 g sugar at the three levels of 2, 4, or 6 g MGTP 100 g$^{-1}$ of flour) were ground, kept overnight in a freezer (−20 °C), and then freeze-dried (Christ Alpha 1-4 LD, SciQuip Ltd., Shropshire, UK) and packed into vacuum-sealed bags until HPLC analysis.

## 2.3. Defatting of Samples and Extraction of Catechins

Freeze-dried sample was weighed (1 g), mixed with 30 mL of hexane, and placed in a heat circulating water bath (Grant Instruments, Cambridge, UK) at 70 °C for 20 min. After cooling down, samples were centrifuged at 3000 rpm for 10 min. The hexane fraction was decanted. The samples were left in the fume cupboard with the light off for 2–3 h to evaporate the hexane from the samples. Defatted samples were extracted by adding 25 mL of 70% methanol with 0.3% formic acid and incubated in a water bath at 70 °C for 5 min to equilibrate to the temperature and another 45 min to extract. After cooling down, the tubes were centrifuged at 3000 rpm for 10 min and the supernatant was collected. The extraction was repeated, the supernatant was pooled together, and its volume was made to 50 mL. The samples were filtered through a 0.2 μm PTFE (Polytetrafluoroethylene) syringe filter and put in amber vials for further HPLC analysis.

## 2.4. HPLC Analysis of Catechins in Dough and Biscuit Samples

A method for HPLC analysis was adapted from Wang and Zhou [17]. The separation was performed on a Phenomenex C18 reverse-phase column (5 μm, 250 × 4.6 mm) (Phenomenex, Cheshire, UK) with a conventional HPLC coupled with a UV-Vis detector (SPD-20A, Shimadzu Corporation, Kyoto, Japan). A binary gradient of water with 0.1% formic acid (eluent A) and acetonitrile with 0.1% formic acid (eluent B) was used to run on the system at a flow rate of 0.5 mL·min$^{-1}$. The gradient started at 10% B, remained at 10% for 5 min, and increased to 15% over 9 min, then climbed to 60% over 23 min, then to 95% over 3 min, and remained at 95% for 2 min to wash the column before returning to 10% over 2 min and re-equilibrating over 6 min. The total length of the method was 45 min. The column temperature was set at 25 °C. Catechins and caffeine were detected at 275 nm. Identification of each catechin was performed by comparing the retention time and spectrum with the standard. Standard curves for quantification were prepared with standard compounds dissolved in extracting solvent (70% methanol with 0.3% formic acid) at concentrations between 2.5–100 μg·mL$^{-1}$. The standards were caffeine, EC, EGCG, ECG, EGC and GCG.

## 2.5. Sensory and Acceptability Studies

Ethical approval for the sensory evaluation study was granted by the Mathematics, Physical Sciences, and Engineering Ethical Committee at the University of Leeds (MEEC 13-026). Sensory evaluation of biscuits was conducted by human panelists, who scored the biscuits using a 9-point hedonic scale. The attributes tested were overall acceptability, appearance, aroma, colour, texture, bitterness, and sweetness. All sensory evaluation sessions were carried out in separate booths equipped with a Compusense 5.6 sensory software (Compusense Inc., Guelph, ON, Canada), in which responses were recorded directly by the participants.

The tests evaluated 9 shortbread formulations with 3 levels of MGTP and 3 levels of sugar. Samples were assigned with 3-digit codes, and the software randomised their serving orders. Plain water was provided to rinse the mouth between samples. The evaluations were repeated in two sessions with same participants. Fifty-four participants completed the first session, and 46 participants completed the second session.

*2.6. Human Intervention Study for Study of Metabolic Response*

Ethical approval for this part of the study was granted by the Mathematics, Physical Sciences, and Engineering Ethical Committee at the University of Leeds (MEEC 14-040). The study was designed according to the FAO (Food and Agriculture Organization) protocol [18]. Healthy adults (*n* = 4 were Asian, *n* = 3 were South American, *n* = 2 were European, *n* = 1 was African) were recruited to the study after they met the eligibility criteria according to a health screening questionnaire (18–60 years; not allergic to any food; not pregnant or lactating; not diagnosed with chronic diseases such as diabetes, cancer, cardiovascular or digestive diseases; not taking any medication that affect the glucose and triglyceride response). The sample size was decided based on harmonized GI methodology [19] who established that 10 individuals is probably enough in most cases to obtain a significant difference between. The design of the study was a cross-over randomised control study without blinding with a one week washout period between sessions. Each participant was asked to attend following overnight fasting overnight for a least 10 h but was not asked to modify their usual diet.

2.6.1. Reference and Test Meals

Three meals (one reference and two test meals) were tested by all panelists in this study in a random order. The reference meal consisted of 100 g of plain shortbread biscuits consumed with 300 mL of warm water. The two test meals consisted of either 100 g of MGTP-enriched biscuits consumed with 300 mL of warm water or 100 g of plain shortbread biscuits consumed with 200 mL of MGTP consumed as a tea drink followed by 100 mL of warm water (total liquid volume of 300 mL). The test meals contained 54–55g of available carbohydrate, of which 16.0 g was sugar and 36 g was fat. The MGTP-enriched biscuit contained 6 g of MGTP (catechin dose of 233 mg). Plain biscuits were prepared according to exactly the same procedure but without addition of MGTP. Once the biscuits were baked and cooled down, 100 g of biscuits were weighed and packed in vacuum-seal plastic bags and used within 1 month for the human study. Matcha green tea drink was made before serving by mixing MGTP (3 g, catechin dose of 257 mg) with 200 mL of warm water. All meals contained around 54–55 g of available carbohydrate. The composition information for the biscuits tested can be found as Supplementary Material (Table S1).

2.6.2. Glucose and Triglyceride Measurement

Capillary blood glucose and triglyceride concentrations were measured from a finger-prick blood sample. A finger was sanitized with an antiseptic wipe before perforation of the skin with a Safe T pro Accu chek disposable lancet. The blood droplet was loaded onto a glucose test strip (Accu-Chek compact 17-drum test strips, Roche diagnostic Ltd., West Sussex, UK) and inserted into a glucometer (Accu-Chek® Aviva blood glucose meter, Roche diagnostic Ltd., West Sussex, UK), which returned the blood glucose concentration in $mmol \cdot L^{-1}$. Another blood droplet was collected by a sterile disposable pipette (15 µL Safetec Pipettes, BHR Pharmaceutical Ltd., Nuneaton, UK). The extracted blood was transferred to a triglyceride test strip (CardioChek® PTS Panel triglyceride test strips, BHR Pharmaceutical Ltd., Nuneaton, UK), which was inserted into a Cardiochek Professional analyser (CardioChek® PA Blood Analyser, BHR Pharmaceutical Ltd., Nuneaton, UK), to measure the blood triglyceride concentration in $mmol \cdot L^{-1}$.

Overnight fasted blood glucose and blood triglycerides were measured at baseline; then, the subject was provided the meal to be consumed within a period of 10 min. Eight blood glucose measurements were taken at 15, 30, 45, 60, 90, 120, 150 and 180 min after the start of the food consumption. Three blood triglyceride measurements were taken at 60, 120 and 180 min after the start of the food consumption.

### 2.6.3. iAUC Calculation

The incremental area under the curve (iAUC) of the glucose and triglyceride response was calculated according to FAO [18], using the trapezoidal rule in which all the areas of glucose response collected during the three hours period are added together, by ignoring the area beneath the baseline.

### 2.7. Data Analysis

Results were analysed statistically to determine mean values, standard deviations (STDs), and standard error of means (SEMs) of quantified masses of compounds obtained from HPLC in duplicate runs. The total catechins content was presented by the sum of the amounts of individual catechins (EGCG, ECG, EGC, GCG and EC). Mann-Whitney U Test was performed to determine the difference between before and after baking of catechin content.

For acceptability sensory results, one factor analysis of variance (ANOVA) with a significant level of $p < 0.05$ and Tukey's-b post-hoc test was performed to determine the difference in scores between the different biscuit formulations. Two factor ANOVA with a significant level of $p < 0.05$ and Tukey's-b post-hoc test was conducted to determine the difference of the acceptability between MGTP and sugar incorporated and their interaction effects. The response surface methodology was conducted using R program (R version 3.2.5, with R Commander package) to plot response surface and contour plot that explained the relationship between the independent variable: MGTP ($X_1$) and sugar ($X_2$); response variables; and testing acceptability attributes (Y): overall appearance, aroma, colour, texture, bitterness, and sweetness of the biscuit samples.

For the human metabolic study, one-way ANOVA and Tukey's-b post hoc test was conducted to investigate the difference between test meals that affects the glucose response and triglyceride response.

Statistical analyses were carried out using the SPSS 22.0 statistical software package (SPSS Inc., Chicago, IL, USA).

## 3. Results and Discussion

### 3.1. Stability of Catechins and Caffeine During the Baking Process

Biscuits with three different levels of MGTP were prepared (see Figure S1). Catechins and caffeine content in dough and biscuits were measured by HPLC-DAD (diode-array detector) (Figure 2, for chromatograms see Figure S2). The results indicate a significant loss of most catechins, except GCG, during the baking process, with a maximum loss of 19% in total catechins (Table 1). EGC presented the highest losses of 31% in biscuits enriched with the highest level of MGTP (6 g 100 g$^{-1}$ of flour). However, there was an increase in GCG of up to 40% in the same biscuits, indicating that epimerization of EGC to GCG may have occurred during the baking process. Caffeine content also decreased during the baking process by up to 24%. The levels of total catechins in the highly enriched biscuits (6 g MGTP 100 g$^{-1}$ of flour) were approximately 20–23 mg per one biscuit, which is around 9% of the amount present in a typical green tea infusion (257 mg 200 mL$^{-1}$ cup).

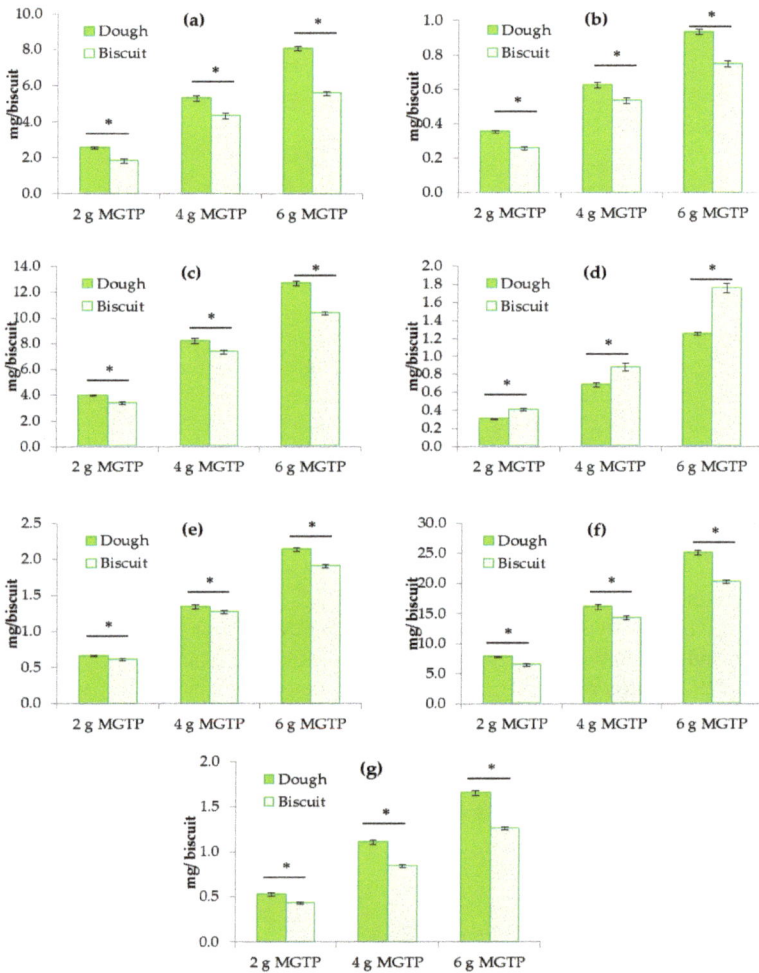

**Figure 2.** Amount of catechins in dough and shortbread enriched with 3 levels of MGTP (2, 4, 6 g 100 g$^{-1}$ of flour): (a) EGC; (b) EC; (c) EGCG; (d) GCG; (e) ECG; (f) total catechins and (g) caffeine. Results are expressed as mg per biscuit. * indicates a significant difference ($p < 0.05$) before and after baking. The error bars represent the SEM (standard error of the mean) ($n = 12$ biscuits).

**Table 1.** Percentage retention of catechins and caffeine in MGTP (Matcha green tea powder)-enriched shortbread biscuits after baking as a proportion of initial content in the dough.

| Compound | 2 g MGTP (%) | 4 g MGTP (%) | 6 g MGTP (%) |
|---|---|---|---|
| (−)-epigallocatechin (EGC) | 70.9 ± 9.5 | 81.5 ± 2.5 | 68.9 ± 1.6 |
| (−)-epicatechin (EC) | 72.6 ± 4.5 | 85.7 ± 3.5 | 80.2 ± 1.7 |
| (−)-epigallocatechin gallate (EGCG) | 84.9 ± 4.0 | 89.7 ± 2.1 | 81.8 ± 1.4 |
| (+)-gallocatechin gallate (GCG) | 133.0 ± 7.3 | 128.6 ± 11.0 | 140.3 ± 8.6 |
| (−)-epicatechin gallate (ECG) | 92.2 ± 3.9 | 94.3 ± 2.4 | 89.1 ± 1.2 |
| Total catechins | 82.4 ± 5.7 | 88.9 ± 2.5 | 81.1 ± 1.4 |
| Caffeine | 81.9 ± 5.0 | 76.3 ± 3.4 | 76.4 ± 1.1 |

The results are in agreement with the study by Wang and Zhou that showed an average retention of 83% for EGCG and 91% for ECG in green tea-enriched bread [17]. Epimerization of EGCG to GCG has been observed in bread [20]. Sharma and Zhou showed much higher losses of catechins during baking (up to 98%) [21]. The degradation followed first order kinetic parameters and could largely be prevented by acidification of the dough. Core temperatures of 120–130 °C achieved during baking would provide sufficient energy for epimerization of EGCG or EGC to GCG [21]. EGC is less stable than other catechins because of its radical scavenging hydroxyl group at the 5′ position of the B ring [22,23]. Therefore, this is could explain the increase in GCG in the MGTP-enriched biscuits.

### 3.2. Sensory and Overall Acceptability of Biscuits

The response surface for overall acceptability (Figure 3a) shows that biscuits with low MGTP and low sugar content received the highest acceptability. The increase in sugar in biscuit formulation did not significantly affect the acceptability of the enriched biscuits. A similar response was observed for appearance (Figure 3b), with darker green and brown colour receiving lower acceptability scores. As shown in Figure 3c, the highest response of aroma acceptability was achieved around a sugar level of 30 g 100 g$^{-1}$ of flour. The surface response method suggested the stationary point at 1.84 g of MGTP and 27.73 g of sugar 100 g$^{-1}$ of flour to obtain the highest response. Heat treatment can cause undesirable aromas of green tea [24]. The trained panelist detected more undesirable (described as wet wood smell) aroma of heat processed green tea at 121 °C for 1 min compared to unprocessed green tea [24]. Therefore, the aroma of shortbread biscuits containing MGTP may be affected by baking temperature at 180 °C, as it will likely cause an undesirable aroma. However, Ahmad et al. [8] found that increasing green tea powder in cookies (1–4%) increased the aroma acceptability.

As shown in Figure 3d, the highest response of colour acceptability was for the biscuits with a sugar level of 30 g 100 g$^{-1}$ of flour and low MGTP. The colour is thought to come from chlorophyll pigments, which can degrade at high temperatures. Kim et al. [25] suggested that green tea solution became less green and deeper yellow after heating at 120 °C for 4 min. Therefore, masking the colour of biscuits by adding fresh green colouring may increase the acceptability of biscuits.

Figure 4a shows that high sugar and low content of MGTP increased the texture acceptability with an optimum score for biscuits with 30 g sugar 100 g$^{-1}$ of flour and low MGTP. Sugar increases the crunchiness of biscuits, while MGTP increases the hardness of biscuits (data is shown in supplementary doc, Figure S3). According to Figure 4b, the bitterness acceptability was significantly affected by the level of MGTP incorporated into the biscuit ($p < 0.001$). The bitterness acceptability received a high response with biscuits that had a low level of MGTP, and sugar did not improve the acceptability. Bitterness can significantly suppress the sense of sweetness and vice versa [26]. It was found that bitterness can be perceived at a much lower concentration than sweetness at a ratio of 1:31 magnitude [27]. Finally, as shown in Figure 4c, sweetness acceptability was significantly affected by the level of MGTP in the biscuit ($p < 0.05$), with higher acceptability for sugar content of 30 g 100 g$^{-1}$ of flour for all MGTP levels tested. This indicates that sweetness does improve the acceptability of biscuits, particularly those with high MGTP. According to the sensory trial, appearance, color, and bitterness were the most important determinants of overall acceptability.

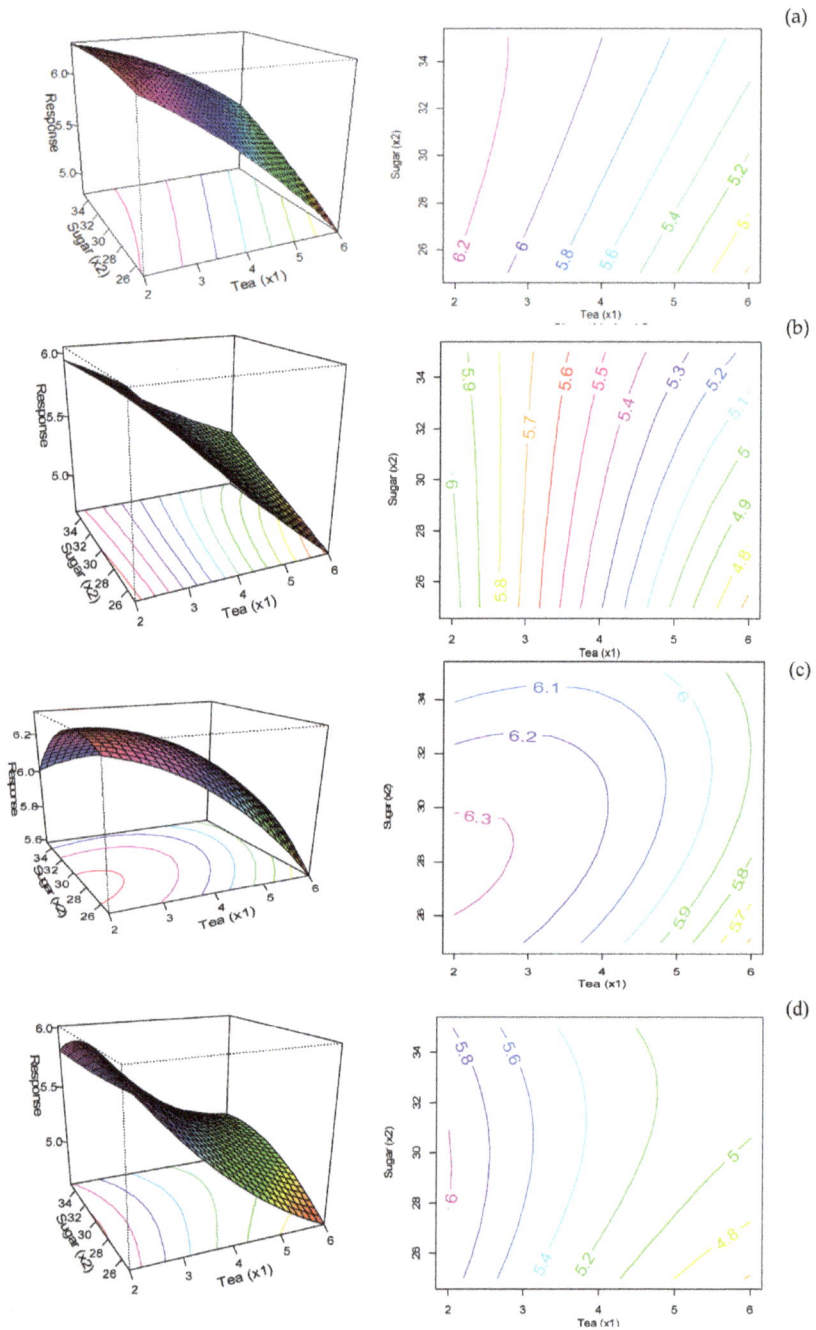

**Figure 3.** Response surfaces and contour plots of the effect of MGTP and sugar content incorporated (**a**) overall; (**b**) appearance; (**c**) aroma and (**d**) color acceptability of biscuits.

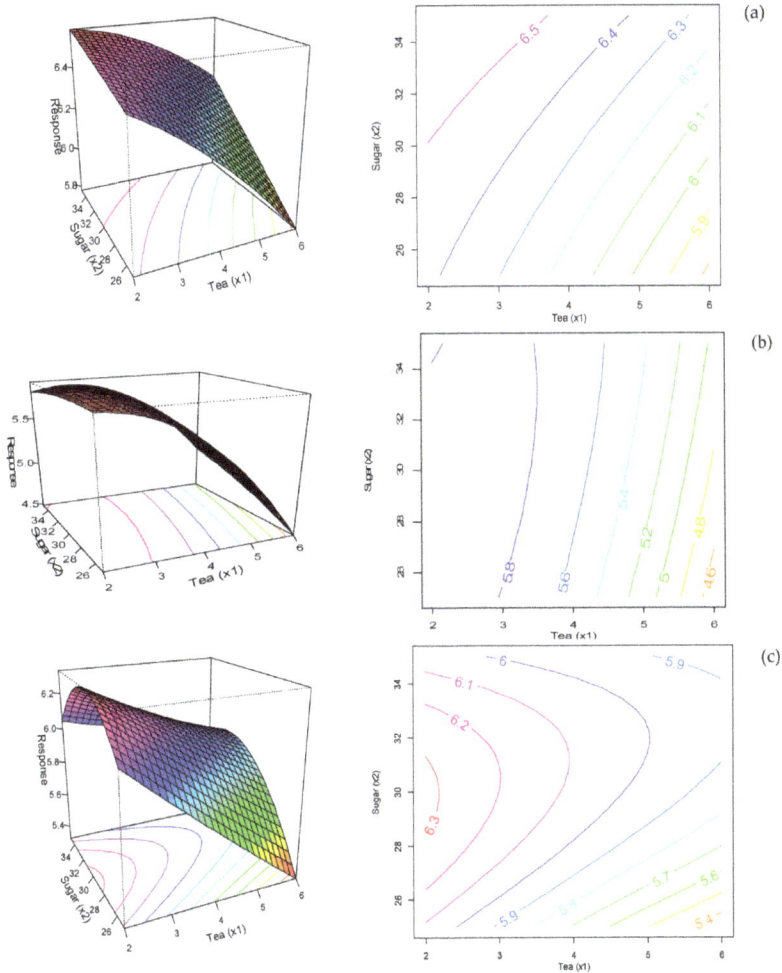

**Figure 4.** Response surfaces and contour plots of the effect of MGTP and sugar content incorporated (**a**) texture; (**b**) bitterness and (**c**) sweetness acceptability of biscuits.

## 3.3. Metabolic Response to Matcha Green Tea Biscuits

In this pilot study, 10 subjects were recruited with ages ranging from 27 to 44 years and a mean BMI of $26.67 \pm 4.48$ kg·m$^{-2}$, with a gender split of 6 females and 4 males. The portions of biscuits served to participants are shown in supplementary material (Figure S4).

### 3.3.1. Glucose Response

Capillary blood glucose concentration was measured at baseline and every 15 min after consumption of the control or test meals. As shown in Figure 5, the glucose response of all of the food samples peaked at 30 min after meal intake, with a slow decrease until 180 min. Consumption of the control meal consisting of plain biscuits gave the highest average iAUC ($106.1 \pm 63.0$ mmol·min L$^{-1}$), followed by the test meal with MGTP-enriched biscuits ($89.2 \pm 44.7$ mmol·min L$^{-1}$) and the test meal with plain biscuits and a MGTP drink ($81.9 \pm 36.0$ mmol·min L$^{-1}$). No significant differences in the iAUCs were observed between meals (one factor ANOVA). Although MGTP does have a tendency to lower iAUC, the variation between individuals is high. The response in the first 30 min appears to be the same amongst the foods, indicating that Matcha green tea does not affect rapidly available glucose (RAG), most probably from sucrose and rapidly digestible starch (RDS). The biggest differences appeared in later time points (>100 min), pointing to a potential inhibition of starch digestion by MGTP, whether in biscuit or tea form. There were significant differences in incremental blood glucose levels at time points 120 and 150 min post ingestion between the plain biscuits and the Matcha treatments.

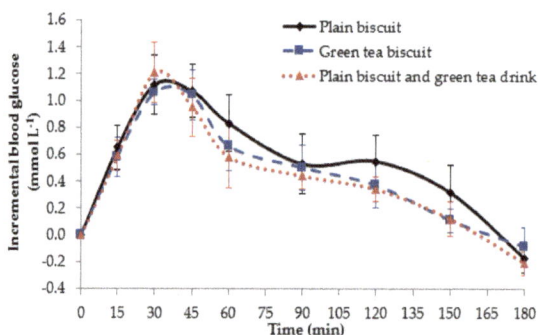

**Figure 5.** Mean incremental blood glucose response curves after ingestion of control and test meals over 180 min. (-◆-) represents the blood glucose curve of the control meal that consisted of plain biscuit consumed with warm water, (-■-) represents the blood glucose curve of the test meal with Matcha green tea biscuits consumed with warm water, and (..▲..) represents the blood glucose curve of the test meal with plain biscuits consumed with a Matcha green tea drink. Data expressed as the amount of blood glucose in mmol·L$^{-1}$ and the error bars represent the SEM ($n = 10$).

Previous findings on the effect of green tea on glucose response have been inconclusive. Many in vivo and in vitro studies have suggested that green tea and EGCG acutely reduce postprandial blood glucose levels by inhibiting the activity of pancreatic α-amylase [28,29]. In an in vitro study, Forester et al. [28] found that EGCG non-competitively inhibited pancreatic amylase activity by 34%. Moreover, Tsuneki et al. [30] showed that glucose tolerance was significantly ($p < 0.05$) improved with tea administration compared with hot water during a glucose challenge in healthy humans. The authors speculated that the observed acute effects of green tea on blood glucose levels were mainly due to the promotion of insulin action in peripheral tissues, such as skeletal muscles and adipocytes via modification of serum protein [30]. Lochocka et al. [31] found that a single dose of green tea extract taken with a test meal significantly decreased starch digestion and absorption compared with the placebo treatment ($p = 0.003$). The study used a $CO_2$ starch breath test, which measures the isotope ratio of $_{12}CO_2/_{13}CO_2$ in breath samples. The starch digestion and absorption was measured based on cumulative percentage $_{13}C$ dose recovery [31]. They found that green tea extract (257.6 mg of catechins) decreased starch digestion and absorption in humans when consumed with a test meal. The green tea extract was used in powder form (4 g of powder containing 257.6 mg of catechins) enclosed in a starch wafer, whereas the placebo was an empty starch wafer. In our study, we observed a significant difference between responses to plain biscuits and the two green tea treatments after 120 and 150 min, suggesting that catechins may have a mild effect on starch digestion. Josic et al. [32] found that green tea infusion (150.6 mg of catechins) showed no effects on glucose or insulin levels. An animal study suggested that EGCG acutely reduces postprandial blood glucose levels in mice when co-administered with corn starch, and this may be due in part to inhibition of α-amylase [28]. Moreover, a study on the effect of green tea supplementation on insulin sensitivity in rats suggested that regular green tea infusion drink could increase insulin sensitivity [33,34]. They also suggested that the amelioration of insulin resistance by green tea is associated with increased expression of glucose transporter IV (GLUT IV) found in adipose tissue. On the other hand, Park et al. [35] proposed that gallated catechins could elevate blood glucose levels by blocking glucose uptake into tissues.

A factor that could have affected glucose response in the present study is the high fat content in shortbread biscuits. The fat content in food can lower the glucose response by slowing down gastric emptying [36]. Hence, the potential differences in starch digestion between control and test meals could have been masked by the high fat content in biscuits. In order to minimize the varied result from sugar and fat content in biscuits, the biscuits with less sugar and low fat content should be used to investigate the effect of MGTP on glucose response. We observed no effect of ethnicity on the glyceamic response to the biscuits. Previous studies have also not identified ethnicity as a significant factor to consider when testing glyceamic response [19].

### 3.3.2. Triglyceride Response

The incremental postprandial responses of plasma triglycerides after consumption of the control and test meals are shown in Figure 6a. When the participants consumed the control meal, plasma triglycerides rose from the baseline to reach the highest level at 120 min and almost returned back to baseline 180 min after consumption of the control meal. The concentration of triglycerides after the consumption of the test meal with Matcha green tea biscuits remained close to the baseline value until 60 min, then rose to a peak at 120 min and returned to baseline at 180 min. The reduction in iAUC after consumption of Matcha green tea biscuits ($35.6 \pm 44.6$ mmol·min L$^{-1}$) was observed, compared with the consumption of plain biscuits ($44.6 \pm 32.1$ mmol·min L$^{-1}$) and plain biscuits consumed with Matcha green tea drink ($39.4 \pm 35.7$ mmol·min L$^{-1}$). However, statistical analysis found that there was no significant difference in iAUC between all three food samples ($p = 0.87$) due to the high variation between individuals.

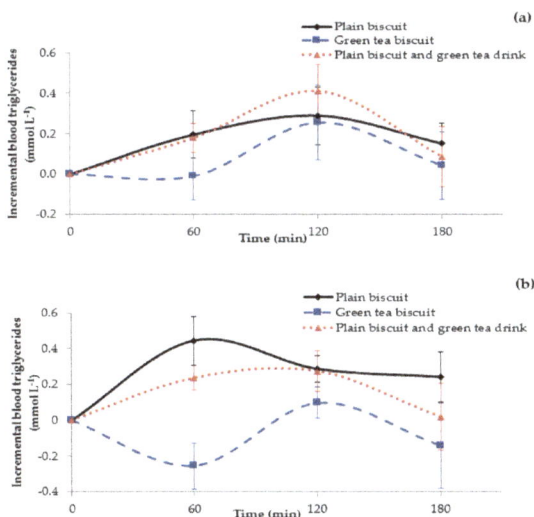

**Figure 6.** (**a**) The mean incremental blood triglyceride response curves after ingestion of control and test meals over 180 min ($n = 10$ participants); (**b**) the mean incremental blood triglyceride response curves after ingestion of control and test meals over 180 min among Asian subjects ($n = 4$ participants). (-◆-) represents the blood triglyceride curve of the control meal consisting of plain biscuits, (-■-) represents the blood triglyceride curve of the test meal with Matcha green tea biscuits, and (..▲..) represents the blood triglyceride curve of the test meal with plain biscuits consumed with a Matcha green tea drink. Data expressed as the amount of triglycerides in mmol L$^{-1}$ and the error bars represent the SEM.

Figure 6b shows the incremental postprandial levels of plasma triglycerides after ingestion of the control and test meals for 4 Asian subjects compared to the other ethnicities. The curves show an average 0.26 mmol·L$^{-1}$ decrease in plasma triglycerides following consumption of the test meal with the MGTP enriched biscuits, whereas the control meal with the plain biscuits induced an average 0.44 mmol·L$^{-1}$ increase in plasma triglycerides. The iAUC of the test meal with Matcha green tea biscuits (8.0 ± 12.5 mmol·min L$^{-1}$) was significantly lower ($p = 0.02$) than the other meals (51.7 ± 22.3 mmol·min L$^{-1}$ and 32.7 ± 17.6 mmol·min L$^{-1}$) for the control meal and test meal (plain biscuits with MGTP tea drink), respectively.

A previous study by Unno et al. [16] found that green tea drink consumed with a piece of bread covered with 20 g of butter lowered the postprandial triglyceride level compared to a meal with a control drink. In that study, there were 2 treatments of green tea; a moderate dose containing 224 mg of tea catechins and a high dose containing 674 mg of tea catechins. Both moderate and high dose treatments lowered the postprandial triglyceride levels and reduced the iAUC by 15.1% and 28.7%, respectively. Only the high dose showed a significant difference in iAUC compared to the control dose. Our findings are in agreement, as the dose of catechins (224 mg of tea catechins) reduced the iAUC by 15.1% but did not give a significant difference. The test meal used in the Unno et al. [16] study was lower in fat (18.8 g of fat) compared to our biscuits (35 g of fat).

The mechanism underlying the triglyceride lowering effect of green tea is mainly focused on the inhibition of the digestion of fat from the intestine [37]. Juhel et al. [38] found that green tea exhibits potential to inhibit gastric and pancreatic lipases. EGCG can enlarge fat droplets through interference with emulsification and micellar solubility of lipids [39–41]. Hence, the enlarged emulsion droplets can prevent the efficient emulsification of bile salt and reduce surface area for fat digestion by lipase [14,16,42,43]. Moreover, Suzuki et al. [37] proposed that catechins with gallate groups are located on the surface of the lipid emulsion and destabilize the lipid emulsion. The difference in responses according to ethnicity merits further investigation. Apolipoproteins are essential regulators of triglyeride circulation. Gene polymorphisms for the a1/c3/a4/a5 gene cluster were found to differ between Chinese and Caucasian populations [44]. Another study found that the postprandial triglyceride concentrations in South Asian men were significantly higher than European men [45]. This was indeed our observation in response to the plain biscuits with a significant effect of the green tea enrichment. These results need to be confirmed in a larger study stratified according to ethnic groups. Information about habitual diet, including catechin content, needs to be considered. Furthermore, more time points in the triglyceride time course need to be included, as triglyceride levels may take around 8 h to return to homeostatic levels.

## 4. Conclusions

This study showed that catechins in MGTP are relatively heat resistant, with only approximately 20% loss of total catechins during baking at 180 °C. Epimerization between EGC and GCG is thought to have occurred during biscuit baking, as the result found a decrease in EGC and an increase in GCG in the baked shortbread biscuits. The effect of MGTP on shortbread biscuit acceptability was assessed using a 9-point hedonic scale and the data was analysed with response surface methodology. With an increase in the level of green tea powder, the acceptability of biscuits decreased significantly, and sugar did not significantly improve acceptability of the enriched biscuits. The main factors affecting acceptability were colour and bitterness. The consumption of MGTP as biscuit or tea infusion did not have significant effects on acute metabolic response, but some interesting trends were observed amongst a low number of Asian participants. This indicates potential for MGTP-enriched products as functional foods. An additional study with larger sample size that takes into account the ethnicity of participants is needed to confirm the observed trends.

**Supplementary Materials:** The following are available online at www.mdpi.com/2304-8158/7/2/17/s1, Figure S1: Plain biscuit (**a**) and green tea biscuits with 3 levels of MGTP incorporated: (**b**) 2 g, (**c**) 4 g and (**d**) 6 g of MGTP 100 g$^{-1}$ of flour, Figure S2: Chromatograph profiles of tea catechins, (**A**) in dough; (**B**) in plain and green tea biscuits: 1. EGC; 2. Caffeine; 3. EC; 4. EGCG; 5. GCG and 6. ECG, Figure S3: Hardness plot of biscuits with 3 levels of MGTP and sugar incorporated. The error bars represent the STDs ($n = 6$), Figure S4: Estimated nutritional data of food samples.

**Acknowledgments:** The authors wish to thank the Ratchabhat Phuket University for funding through Ph.D. sponsorship to BP.

**Author Contributions:** B.P., C.O., M.H. and L.J.M. conceived and designed the experiments; B.P. performed the experiments; B.P., C.O., M.H. and L.J.M. analysed the data, contributed reagents/materials/analysis tools, and wrote the paper.

**Conflicts of Interest:** The authors declare no conflict of interest.

## References

1. Wang, H.; Provan, G.; Helliwell, K.; Ransom, W. The functional benefits of flavonoids: The case of tea. In *Phytochemical Functional Foods*; Johnson, I., Williamson, G., Eds.; Woodhead Publishing: Sawston, Cambridge, UK, 2003; pp. 128–159.

2. Wang, R.; Zhou, W.; Jiang, X. Reaction kinetics of degradation and epimerization of epigallocatechin gallate (EGCG) in aqueous system over a wide temperature range. *J. Agric. Food Chem.* **2008**, *56*, 694–2701. [CrossRef] [PubMed]

3. Pelillo, M.; Biguzzi, B.; Bendini, A.; Toschi, T.G.; Vanzini, M.; Lercker, G. Preliminary investigation into development of HPLC with UV and MS-electrospray detection for the analysis of tea catechins. *Food Chem.* **2002**, *78*, 369–374. [CrossRef]

4. Mennen, L.; Hirvonen, T.; Arnault, N.; Bertrais, S.; Galan, P.; Hercberg, S. Consumption of black, green and herbal tea and iron status in French adults. *Eur. J. Clin. Nutr.* **2007**, *61*, 1174–1179. [CrossRef] [PubMed]

5. Kim, Y.-K.; Jombart, L.; Valentin, D.; Kim, K.O. A cross-cultural study using Napping®: Do Korean and French consumers perceive various green tea products differently? *Food Res. Int.* **2013**, *53*, 534–542. [CrossRef]

6. Šebečić, B.; Vedrina Dragojević, I.; Vitali, D.; Hečimović, M.; Dragičević, I. Raw materials in fibre enriched biscuits production as source of total phenols. *Agric. Conspec. Sci.* **2007**, *72*, 265–270.

7. Carrillo, E.; Varela, P.; Fiszman, S. Effects of food package information and sensory characteristics on the perception of healthiness and the acceptability of enriched biscuits. *Food Res. Int.* **2012**, *48*, 209–216. [CrossRef]

8. Ahmad, M.; Baba, W.N.; Wani, T.A.; Gani, A.; Gani, A.; Shah, U.; Wani, S.M.; Masoodi, F.A. Effect of green tea powder on thermal, rheological & functional properties of wheat flour and physical, nutraceutical & sensory analysis of cookies. *J. Food Sci. Technol.* **2015**, *52*, 5799–5807. [PubMed]

9. Lu, T.M.; Lee, C.C.; Mau, J.L.; Lin, S.D. Quality and antioxidant property of green tea sponge cake. *Food Chem.* **2010**, *119*, 1090–1095. [CrossRef]

10. Li, M.; Zhang, J.H.; Zhu, K.X.; Peng, W.; Zhang, S.K.; Wang, B.; Zhu, Y.J.; Zhou, H.M. Effect of superfine green tea powder on the thermodynamic, rheological and fresh noodle making properties of wheat flour. *LWT Food Sci. Technol.* **2012**, *46*, 23–28. [CrossRef]

11. Yang, J.; Mao, Q.X.; Xu, H.X.; Ma, X.; Zeng, C.Y. Tea consumption and risk of type 2 diabetes mellitus: A systematic review and meta-analysis update. *BMJ Open* **2014**, *4*, e005632. [CrossRef] [PubMed]

12. Thielecke, F.; Boschmann, M. The potential role of green tea catechins in the prevention of the metabolic syndrome—A review. *Phytochemistry* **2009**, *70*, 11–24. [CrossRef] [PubMed]

13. Zhong, L.; Furne, J.K.; Levitt, M.D. An extract of black, green, and mulberry teas causes malabsorption of carbohydrate but not of triacylglycerol in healthy volunteers. *Am. J. Clin. Nutr.* **2006**, *84*, 551–555. [PubMed]

14. Koo, S.I.; Noh, S.K. Green tea as inhibitor of the intestinal absorption of lipids: Potential Mechanism for its Lipid-Lowering Effect. *J. Nutr. Biochem.* **2007**, *18*, 179–183. [CrossRef] [PubMed]

15. Ikeda, I.; Tsuda, K.; Suzuki, Y.; Kobayashi, M.; Unno, T.; Tomoyori, H.; Goto, H.; Kawata, Y.; Imaizumi, K.; Nozawa, A.; et al. Tea catechins with a galloyl moiety suppress postprandial hypertriacylglycerolemia by delaying lymphatic transport of dietary fat in rats. *J. Nutr.* **2005**, *135*, 155–159. [CrossRef] [PubMed]

16. Unno, T.; Tago, M.; Suzuki, Y.; Nozawa, A.; Sagesaka, Y.M.; Kakuda, T.; Egawa, K.; Kondo, K. Effect of tea catechins on postprandial plasma lipid responses in human subjects. *Br. J. Nutr.* **2005**, *93*, 543–547. [CrossRef] [PubMed]

17. Wang, R.; Zhou, W. Stability of tea catechins in the breadmaking Process. *J. Agric. Food Chem.* **2004**, *52*, 8224–8229. [CrossRef] [PubMed]
18. Food and Agriculture Organization. *Carbohydrates in Human Nutrition. Report of a Joint FAO/WHO Expert Consultation*; FAO Food Nutrition Paper; Food and Agriculture Organization of the United Nations: Rome, Itlay, 1998; Volume 66, pp. 1–140.
19. Brouns, F.; Bjorck, I.; Frayn, K.N.; Gibbs, A.L.; Lang, V.; Slama, G.; Wolever, T.M.S. Glycaemic index methodology. *Nutr. Res. Rev.* **2005**, *18*, 145–171. [CrossRef] [PubMed]
20. Wang, R.; Zhou, W.; Jiang, X. Mathematical modeling of the stability of green tea catechin epigallocatechin gallate (EGCG) during bread baking. *J. Food Eng.* **2008**, *87*, 505–513. [CrossRef]
21. Sharma, A.; Zhou, W. A stability study of green tea catechins during the biscuit making process. *Food Chem.* **2011**, *126*, 568–573. [CrossRef]
22. Guo, Q.; Zhao, B.; Shen, S.; Hou, J.; Hu, J.; Xin, W. ESR study on the structure–antioxidant activity relationship of tea catechins and their epimers. *Biochim. Biophys. Acta (BBA) Gen. Subj.* **1999**, *1427*, 13–23. [CrossRef]
23. Ananingsih, V.K.; Sharma, A.; Zhou, W. Green tea catechins during food processing and storage: A review on stability and detection. *Food Res. Int.* **2013**, *50*, 469–479. [CrossRef]
24. Wang, L.-F.; So, S.; Baik, J.H.; Kim, H.J.; Moon, K.S.; Park, S.K. Aroma changes in green tea beverage during processing and storage. In *Nutraceutical Beverages*; American Chemical Society: Washington, DC, USA, 2003; pp. 162–188.
25. Kim, E.S.; Liang, Y.R.; Jin, J.; Sun, Q.F.; Lu, J.L.; Du, Y.Y.; Lin, C. Impact of heating on chemical compositions of green tea liquor. *Food Chem.* **2007**, *103*, 1263–1267. [CrossRef]
26. Wang, R.; Zhou, W.; Isabelle, M. Comparison study of the effect of green tea extract (GTE) on the quality of bread by instrumental analysis and sensory evaluation. *Food Res. Int.* **2007**, *40*, 470–479. [CrossRef]
27. Green, B.G.; Lim, J.; Osterhoff, F.; Blacher, K.; Nachtigal, D. Taste mixture interactions: Suppression, additivity, and the predominance of sweetness. *Physiol. Behave.* **2010**, *101*, 731–737. [CrossRef] [PubMed]
28. Forester, S.C.; Gu, Y.; Lambert, J.D. Inhibition of starch digestion by the green tea polyphenol, (−)-epigallocatechin-3-gallate. *Mol. Nutr. Food Res.* **2012**, *56*, 1647–1654. [CrossRef] [PubMed]
29. Chen, N.; Bezzina, R.; Hinch, E.; Lewandowski, P.A.; Cameron-Smith, D.; Mathai, M.L.; Jois, M.; Sinclair, A.J.; Begg, D.P.; Wark, J.D.; et al. Green tea, black tea, and epigallocatechin modify body composition, improve glucose tolerance, and differentially alter metabolic gene expression in rats fed a high-fat diet. *Nutr. Res.* **2009**, *29*, 784–793. [CrossRef] [PubMed]
30. Tsuneki, H.; Ishizuka, M.; Terasawa, M.; Wu, J.B.; Sasaoka, T.; Kimura, I. Effect of green tea on blood glucose levels and serum proteomic patterns in diabetic (db/db) mice and on glucose metabolism in healthy humans. *BMC Pharmacol.* **2004**, *4*, 18. [CrossRef] [PubMed]
31. Lochocka, K.; Bajerska, J.; Glapa, A.; Fidler-Witon, E.; Nowak, J.K.; Szczapa, T.; Grebowiec, P.; Lisowska, A.; Walkowiak, J. Green tea extract decreases starch digestion and absorption from a test meal in humans: A randomized, placebo-controlled crossover study. *Sci. Rep.* **2015**, *5*, 12015. [CrossRef] [PubMed]
32. Josic, J.; Olsson, A.T.; Wickeberg, J.; Lindstedt, S.; Hlebowicz, J. Does green tea affect postprandial glucose, insulin and satiety in healthy subjects: A randomized controlled trial. *Nutr. J.* **2010**, *9*, 63. [CrossRef] [PubMed]
33. Wu, L.Y.; Juan, C.C.; Ho, L.T.; Hsu, Y.P.; Hwang, L.S. Effect of green tea supplementation on insulin sensitivity in Sprague-Dawley rats. *J. Agric. Food Chem.* **2004**, *52*, 643–648. [CrossRef] [PubMed]
34. Wu, L.Y.; Juan, C.C.; Hwang, L.S.; Hsu, Y.P.; Ho, P.H.; Ho, L.T. Green tea supplementation ameliorates insulin resistance and increase glucose transporter IV content in a fructose-fed rat model. *Eur. J. Nutr.* **2004**, *43*, 116–124. [CrossRef] [PubMed]
35. Park, J.H.; Jin, J.Y.; Baek, W.K.; Park, S.H.; Sung, H.Y.; Kim, Y.K.; Lee, J.; Song, D.K. Ambivalent role of gallated catechins in glucose tolerance in humans: A novel insight into non-absorbable gallated catechin-derived inhibitors of glucose absorption. *J. Physiol. Pharmacol.* **2009**, *60*, 101–109. [PubMed]
36. Wolever, T.M.S. *The Glycaemic Index: A Physiological Classification of Dietary Carbohydrate*; CABI Pub.: Wallingford, Oxfordshire, UK; Cambridge, MA, USA, 2006; 227p.
37. Suzuki, Y.; Unno, T.; Kobayashi, M.; Nozawa, A.; Sagesaka, Y.; Kakuda, T. Dose-dependent suppression of tea catechins with a galloyl moiety on postprandial hypertriglyceridemia in rats. *Biosci. Biotechnol. Biochem.* **2005**, *69*, 1288–1291. [CrossRef] [PubMed]

38. Juhel, C.; Armand, M.; Pafumi, Y.; Rosier, C.; Vandermander, J.; Lairon, D. Green tea extract (AR25®) inhibits lipolysis of triglycerides in gastric and duodenal medium in vitro. *J. Nutr. Biochem.* **2000**, *11*, 45–51. [CrossRef]

39. Kim, J.; Koo, S.I.; Noh, S.K. Green tea extract markedly lowers the lymphatic absorption and increases the biliary secretion of C-14-benzo a pyrene in rats. *J. Nutr. Biochem.* **2012**, *23*, 1007–1011. [CrossRef] [PubMed]

40. Chan, C.Y.; Wei, L.; Castro-Muñozledo, F.; Koo, W.L. (−)-Epigallocatechin-3-gallate blocks 3T3-L1 adipose conversion by inhibition of cell proliferation and suppression of adipose phenotype expression. *Life Sci.* **2011**, *89*, 779–785. [CrossRef] [PubMed]

41. Lee, M.-S.; Kim, C.T.; Kim, I.H.; Kim, Y. Inhibitory Effects of Green Tea Catechin on the Lipid Accumulation in 3T3-L1 Adipocytes. *Phytother. Res.* **2009**, *23*, 1088–1091. [CrossRef] [PubMed]

42. Walkowiak, J.; Bajerska, J.; Kargulewicz, A.; Lisowska, A.; Siedlerski, G.; Szczapa, T.; Kobelska-Dubiel, N.; Grzymisławski, M. Single dose of green tea extract decreases lipid digestion and absorption from a test meal in humans. *Acta Biochim. Pol.* **2013**, *60*, 481–483. [PubMed]

43. Raederstorff, D.G.; Schlachter, M.F.; Elste, V.; Weber, P. Effect of EGCG on lipid absorption and plasma lipid levels in rats. *J. Nutr. Biochem.* **2003**, *14*, 326–332. [CrossRef]

44. Liu, Z.-K.; Hu, M.; Baum, L.; Thomas, G.N.; Tomlinson, B. Associations of polymorphisms in the apolipoprotein A1/C3/A4/A5 gene cluster with familial combined hyperlipidaemia in Hong Kong Chinese. *Atherosclerosis* **2010**, *208*, 427–432. [CrossRef] [PubMed]

45. Arjunan, S.P.; Bishop, N.; Reischak-Oliveira, A.; Stensel, D.J. Exercise and coronary heart disease risk markers in South Asian and European men. *Med. Sci. Sports Exerc.* **2013**, *45*, 1261–1268. [CrossRef] [PubMed]

*foods*

MDPI

Article

# Effects of Two Doses of Curry Prepared with Mixed Spices on Postprandial Ghrelin and Subjective Appetite Responses—A Randomized Controlled Crossover Trial

Sumanto Haldar [1], Joseph Lim [1], Siok Ching Chia [1], Shalini Ponnalagu [1] and Christiani Jeyakumar Henry [1,2,*]

[1]   Clinical Nutrition Research Centre (CNRC), Singapore Institute for Clinical Sciences (SICS), Agency for Science Technology and Research (A*STAR), 30 Medical Drive, Singapore 117609, Singapore; sumanto_haldar@sics.a-star.edu.sg (S.H.); joseph_lim@sics.a-star.edu.sg (J.L.); smileychiasc@gmail.com (S.C.C.); Shalini_Ponnalagu@sics.a-star.edu.sg (S.P.)
[2]   Department of Biochemistry, National University of Singapore, Singapore 119077, Singapore
*   Correspondence: jeya_henry@sics.a-star.edu.sg; Tel.: +65-64-070-793

Received: 23 February 2018; Accepted: 25 March 2018; Published: 26 March 2018

**Abstract:** Spices are known to provide orosensory stimulation that can potentially influence palatability, appetite, and energy balance. Previous studies with individual spices have shown divergent effects on appetite and energy intake measures. In a real-life context, however, several spices are consumed in combinations, as in various forms of curries. Therefore, we investigated changes in postprandial appetite and plasma ghrelin in response to the intake of two doses of curry prepared with mixed spices. The study was undertaken in healthy Chinese men, between 21 and 40 years of age and body mass index $\leq 27.5$ kg/m$^2$. Appetite was measured using visual analogue scales (VAS) and plasma ghrelin was measured using multiplex assay. Compared with the control meal (Dose 0 Control (D0C), 0 g mixed spices), we found significantly greater suppression in 'hunger' (both $p < 0.05$, after Bonferroni adjustments) as well in 'desire to eat' (both $p < 0.01$) during the Dose 1 Curry (D1C, 6 g mixed spices) and Dose 2 Curry (D2C, 12 g mixed spices) intake. There were no differences, however, in plasma ghrelin or in other appetite measures such as in 'fullness' or in 'prospective eating' scores. Overall, the results of our study indicate greater inter-meal satiety due to mixed spices consumption, independent of any changes in plasma ghrelin response.

**Keywords:** spices; curry; ghrelin; appetite response

---

## 1. Introduction

Spices and related flavor compounds consumed worldwide are known to provide orosensory stimulation which can potentially influence palatability and appetite. This in turn can modulate ingestive behavior within meals and between meals and thereby have the potential to alter energy balance [1]. It is an established fact that taste and smell can influence satiety and hunger responses [2,3], although the associations can be divergent, depending on the individual food types [4], as well as on the optimal dosing of sensory intensity [5]. The literature on the intake of individual spices *per se* indicate that the associations between sensory, appetite, and energy intake to be rather equivocal, with some studies showing an increased palatability/liking of foods when spices are added to them [6], whereas other studies finding no such differences [7]. Similarly, regarding appetite response, the findings have been rather variable, with some studies showing no differences in appetite ratings when pepper, ginger, horseradish, etc. were individually added to a mixed dish [8],whereas a recent animal study found appetite enhancing effects of essential oils from both cinnamon and ginger [9]. Similarly, increased use of spices have also been reported in individuals with compromised chemosensory function in order to compensate for the loss of appetite [10], supporting the appetite modulating ability of spices.

In the recently completed 'Polyspice study' we found significant improvements in postprandial glucose homeostasis as well as increases in glucagon like peptide- (GLP-1) response [11,12]. Given the equivocal nature of findings in the current literature regarding spice intake and appetite response, we further explored whether the subjective appetite response will differ between two doses of curry, made with polyphenol rich mixed spices, in comparison with non-curry control within the same Polyspice study cohort. Furthermore, while it is reasonably well established that postprandial ghrelin response can differ depending on the macronutrient composition of the meals [13,14], less is known regarding the effects of dietary bioactive phytochemicals on ghrelin response. Evidence is beginning to emerge that polyphenol content of foods may influence postprandial ghrelin response [15–17], with different polyphenols suggested to have divergent effects [18]. Therefore, we additionally investigated the postprandial ghrelin response to two different doses of curry made with polyphenol rich mixed spices and vegetables. To the best of our knowledge, this is the first dose-response study of its kind exploring the effects of dietary relevant doses of curry intake, made with mixed spices, on postprandial appetite and ghrelin response.

## 2. Materials and Methods

The details of the study design have been described previously [11]. In brief, the study was undertaken in healthy Chinese men between 21 and 40 years of age and body mass index $\leq$27.5 kg/m$^2$. The study was approved by the Domain Specific Research Board (DSRB) ethics committee, Singapore (Reference: C/2015/00729) and was registered on clinicaltrials.gov (Identifier No. NCT02599272). This was a 3-way randomized crossover trial, with each volunteer completing three separate study sessions: Dose 0 Control (D0C), or Dose 1 Curry (D1C), or Dose 2 Curry (D2C) treatments. Twenty volunteers completed D0C and D2C sessions, whereas 17 volunteers completed the D1C (optional) session. During the individual study sessions, each volunteer consumed the test meals for breakfast, after an overnight fast following a standardized dinner the evening before. D0C meal contained no (0 g) spices, D1C meal consisted 6 g of mixed spices, and D2C meal consisted of 12 g mixed spices. The mixed spices preparations for D1C and D2C were identical and were prepared by thoroughly mixing dried powders of different spices consisting of turmeric, coriander seeds, cumin seeds (all Everest Spices, Mumbai, India), dried Indian gooseberry ('amla', *emblica officinalis*, Ramdev Spices, Ahmedabad, India), cayenne pepper (Robertson's, Durban, South Africa), cinnamon (McCormick's, Baltimore, MD, USA), and clove (Robertson's, Durban, South Africa) and were mixed in the ratio of 8:4:4:4:2:1:1, respectively. Test meals were consumed with a portion of white rice, providing a total of approximately 100 g available carbohydrates each and were balanced for total energy, protein, fat, dietary fiber, and total vegetables content. The mean total energy contents for each test meal was approximately 605 kilocalories consisting around 67%, 7%, and 27% energy from carbohydrate, protein, and fat, respectively. Volunteers were asked to consume their meals within 15 min of serving.

Visual analogue scales (VAS) were used to rate subjective appetite sensations as validated previously [19,20]. The VAS consisted of 100 mm long horizontal lines with two ends describing "extremely ... " (coded as 100 mm) and "not at all ... " (coded as 0 mm) with each scale measuring 'hunger', 'fullness', 'desire to eat', and 'prospective eating'. Volunteers were asked to capture their appetite sensations during various times, at regular intervals, by putting vertical mark along the four horizontal lines. These measurements were undertaken immediately prior to the consumption of test meals (baseline, 0 h) followed by 7 postprandial time points at various regular intervals (0.25 h, 0.5 h, 1.0 h, 1.5 h, 2.0 h, 2.5 h, 3 h). Blood samples were also obtained for plasma ghrelin measurements via an intravenous cannula collected during the same times as the appetite response measurements (except for 0.25 h time point). Blood samples were collected in 2 mL K$_2$ Ethylenediaminetetraacetic acid (EDTA) vacutainer tube (BD, Franklin Lakes, NJ, USA) pre-treated with cOmplete™, Mini, EDTA-free protease inhibitor cocktail tablet (Roche, Basel, Switzerland), stored cool on ice, and were centrifuged within 45 min of collection at 1500× $g$ for 10 min at 4 °C. Plasma samples were then immediately stored at −80 °C until analyses. Luminex® bead-based multiplex assays, based on Luminex® xMAP® technology, were used to measure total ghrelin concentration in plasma according to manufacturer's protocol (ProcartaPlex, Thermo Fisher Scientific, Waltham, MA, USA).

Statistical analysis was performed using SPSS, version 24 (IBM Inc., Armonk, North Castle, NY, USA). Data were analyzed using the mixed effects model with the doses as the fixed effect using a compound symmetry covariance structure to test for the overall effect of the doses. Change from baseline response for plasma ghrelin and appetite measures were calculated as the change from baseline areas under the curve (ΔAUC). Post-hoc pairwise comparisons, using Bonferroni corrections, were used to compare differences in change from baseline AUCs between various doses. Furthermore, to test the overall effect of the doses on total AUC data for plasma ghrelin, the corresponding baseline values (fasting values) of the subjects at each of the doses were added as a covariate, with the doses as the fixed effects using a compound symmetry covariance structure. However, for the overall tests of the change from baseline AUCs, no covariates were used. Square root transformations of the data were undertaken to achieve normal distribution, where necessary.

## 3. Results

There were no reported adverse reactions to the test meals, and the volunteers consumed their served meal portions in full within the suggested time allocated, indicating a satisfactory tolerance. Postprandial ghrelin response to the three test meals is shown in Figure 1. There were no significant differences in either the change from baseline AUC (ΔAUC) or in the total areas under the curve (total AUC) between the D0C, D1C, and D2C meals. The mean change from baseline (0 h) of subjective appetite ratings including 'hunger', 'fullness', 'desire to eat', and 'prospective eating' are shown in Figure 2. Compared with the control (D0C) test meal, D1C and D2C led to significantly greater suppressions in 'hunger' (by approximately 54% and 51% during D1C and D2C, respectively, as compared with D0C control meal, both $p < 0.05$, after Bonferroni corrections) as well as in the 'desire to eat' (by approximately 62% and 60% in D1C and D2C, respectively, as compared with D0C control meal, both $p < 0.01$, after Bonferroni corrections), as calculated using the areas under the curve of the change from baseline measurements. The summary data are shown in Table 1. Moreover, even though the mean change from baseline AUC (ΔAUC) of postprandial 'fullness' was also greater by about 20% during D1C and D2C meals as compared with the D0C meal, indicating increased 'fullness', none of the differences reached statistical significance. Similarly, there were no statistical differences in the 'prospective eating' rating between test meals.

**Table 1.** Postprandial changes from baseline (ΔAUC) over 3 h periods in plasma ghrelin concentration and appetite response measures during three test meals (D0C, D1C, and D2C).

| Measurement | D0C (Mean ± SD) (n = 20) | D1C (Mean ± SD) (n = 17) | D2C (Mean ± SD) (n = 20) | Pairwise Comparison * |
|---|---|---|---|---|
| Total Ghrelin (ΔAUC) | −78,098.98 ± 41,101.98 | −76,549.46 ± 41,482.70 | −78,883.94 ± 40,022.88 | ND |
| 'Hunger' (ΔAUC) | −4659.45 ± 3272.36 | −7179.61 ± 3782.34 | −7020.96 ± 3871.20 | D0C vs. D1C ($p = 0.017$) D0C vs. D2C ($p = 0.028$) |
| 'Fullness' (ΔAUC) | 6393.71 ± 3681.80 | 7764.54 ± 3908.07 | 7850.15 ± 3581.19 | ND |
| 'Desire to Eat' (ΔAUC) | −4495.00 ± 3194.31 | −7268.09 ± 4053.52 | −7194.88 ± 3849.50 | D0C vs. D1C ($p = 0.002$) D0C vs. D2C ($p = 0.005$) |
| 'Prospective eating' (ΔAUC) | −4913.78 ± 3401.05 | −6095.76 ± 3612.99 | −5883.42 ± 3506.46 | ND |

* Pairwise comparisons after Bonferroni correction. Δ AUC—changes from baseline area under the curve, ND—no significant difference, D0C—Dose 0 Control, D1C—Dose 1 Curry, D2C—Dose 2 Curry (D2C—), SD—standard deviation.

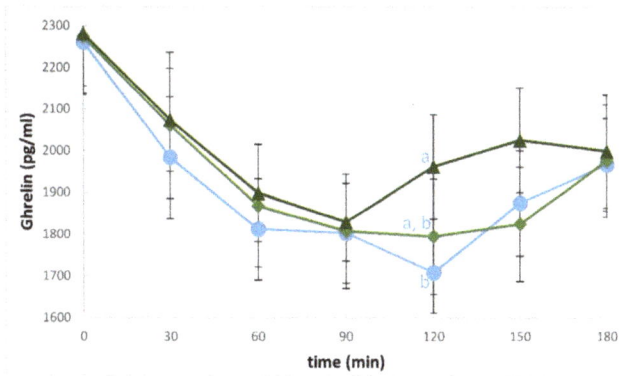

**Figure 1.** Mean (±Standard Error of Mean) postprandial plasma ghrelin concentration in response to the intake of three test meals: ●Dose 0 Control (D0C), ◆ Dose 1 Curry (D1C), ▲ Dose 2 Curry (D2C). The differences between the treatments at each time point were measured after controlling for the baseline values. Treatments that were significantly different from each other ($p$-value < 0.05) are represented by different letters. At time 120 min, there was a significant main effect of the treatment, although there was only marginal significance observed between Dose 0 and Dose 2 ($p$-value = 0.056), as shown in the figure above.

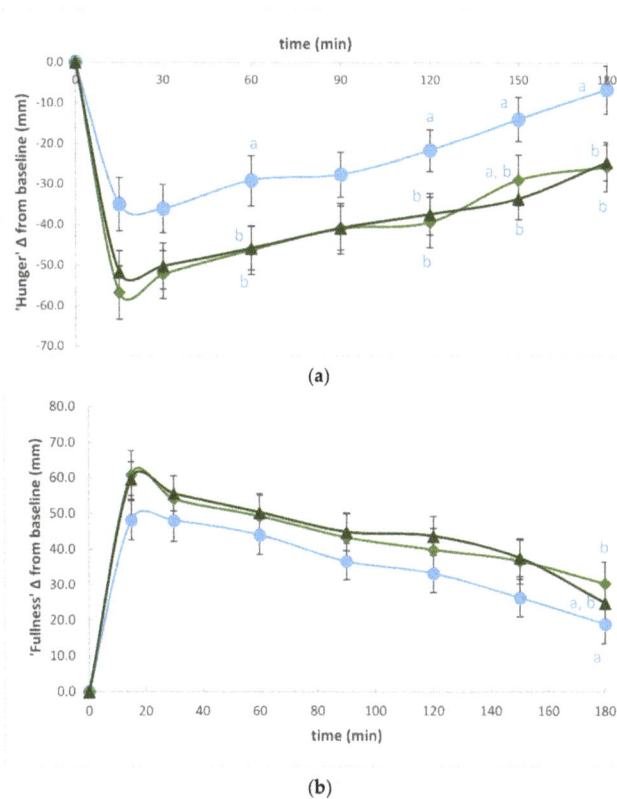

(a)

(b)

**Figure 2.** *Cont.*

(c)

(d)

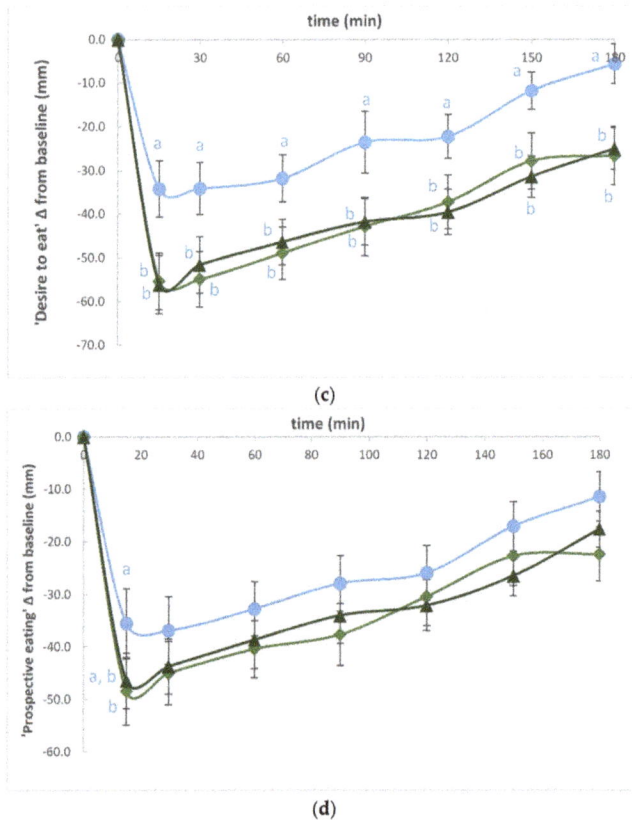

**Figure 2.** Mean (±SEM) in subjective appetite ratings over a 3 h period following consumption of three test meals. (a) 'Hunger'; (b) 'Fullness'; (c) 'Desire to eat'; (d) 'Prospective eating'. ● Dose 0 Control (D0C), ◆ Dose 1 Curry (D1C), ▲ Dose 2 Curry (D2C). In all the appetite ratings, the main effect of time and treatment were significantly different ($p$-value < 0.05). Treatments that were significantly different from each other ($p$-value < 0.05) are represented by different letters at each given time point.

## 4. Discussion

It is generally believed that glucose homeostasis is associated with appetite and/or incretin hormone response and vice-versa [21–23]. While several studies have investigated these effects simultaneously by modulating macronutrient compositions of meals, a limited number of studies have explored the effects of mixed spices in humans using a controlled dose-response trial [1]. Given that we have previously shown beneficial effects of curry made with mixed spices on glucose homeostasis [11], in this additional investigation, we wanted to explore the influence of dose-dependent increases in the intake of curry made with mixed spices on appetite and ghrelin responses. The strengths of our study design were the use of dietary relevant doses of mixed spices that are typically consumed in Indian curries and that the study was deliberately undertaken in a Chinese population, who would not usually consume large amounts of Indian curries, thereby avoiding any potential residual effects due to habituation. Moreover, we also balanced the total energy, macronutrients, and total vegetable contents across the three test meals.

Despite all the test meals being well tolerated, we found significant increases in 'hunger' suppression and in the suppression of 'desire to eat' with the mixed spice containing curry doses as compared with the control meal. This indicates that the consumption of spices may increase inter-meal satiety, which ties-in with our previously reported finding within the same study of an increase in plasma GLP-1 concentration with increasing curry doses [12], which can partly lead to appetite suppression. Other studies have also shown that consumption of individual spices can lead to increases in plasma GLP-1 [24] as well as in peptide YY (PYY) [25] concentrations. Furthermore, the lack of any difference in postprandial ghrelin response between the various test meals in our study is similar to the observation made with two doses of cinnamon added to 300 g of rice pudding [24]. The consistencies in these findings may suggest that the appetite suppression effects of spices may be via the increases in the in vivo concentrations of anorexigenic gut hormones such as GLP-1 and PYY, rather than via an increased suppression of the orexigenic hormone ghrelin. In further support of our findings, a recent study also found appetite suppressing effects of individual spices (e.g., turmeric, cinnamon, ginger) [25]. All these spices (and more) were used together in the mixed spice containing test meals in our study. In contrast to our findings, however, another study investigating the effects of individual spices such as mustard, horseradish, black pepper, and ginger found negligible effects on appetite response [8]. Therefore, different spices are likely to give rise to divergent effects. As such, spices such as chili have been shown to demonstrate appetite stimulating effects [26]. The 'appetizer effects' of spices have also been previously shown by adding them in low salt-foods [27] as well as to reduced fat foods [6]. The contrasting effects of different spices on appetite could be resulting from the diverse ways by which individual spices can affect the sensory and molecular pathways, as have been reviewed in detail elsewhere [1].

Within our study population we found a modest inter-individual variability in the subjective appetite ratings as a result of mixed spice consumption at the various doses, as observed through the large standard deviations in appetite rating measures (Table 1). This could partly be due to differences in prior familiarities and likings, since it has been shown in a recent study that familiarity of spices can determine pleasantness response which in turn can affect appetite [28]. Indeed, prior preference for an individual to sweet or savory foods has been shown to determine the quantity of either sweet or savory foods that is eaten in an experimental setting [29]. There are also indications that behavioral variables including risk taking and different personality traits could also influence the motivation to consume spicy foods [30], and although this previous study was undertaken within the Western dietary context, we used Chinese males who may also have different affiliations towards eating Indian curries. In our study, however, we did not gather any information on prior familiarity, liking, or on personally traits which could directly or indirectly explain part of the inter-individual variabilities in appetite responses.

We did not find any obvious dose-response associations with appetite measures in our study, since both the curry doses seemed to have exerted similar effects as compared with the non-curry control meal. This was despite previous studies suggesting an inverted U-shaped (Wundt) curve between sensory intensity and appetite response as discussed in more detail by McCrickerd et al. [5]. More specifically, both palatability [31] as well as appetite in response to dose-dependent increases in flavor/taste intensities observed this inverted U-shaped phenomenon by several investigators [32,33], although we did not find such effects. These differences in findings could be explained by the fact that the "preferred" taste intensity for spices could have varied between individuals, which may have contributed to the moderate inter-individual variability in appetite response seen in our study. This in turn could lead to lack of any obvious dose-response effects. Moreover, we measured neither the volunteers' innate taste preference nor the palatability of the test meals, which were some of the limitations of this study. Additionally, the volunteers were required to consume the entire amount of test meals served, irrespective of their within meal satiation. Furthermore, the study could have been more objective if we provided them an *ad libitum* meal for lunch and measured the volunteers' actual food intake rather than solely measuring subjective appetite response between meals. Since the primary objective of our study was to investigate metabolic response to standardized, fixed amounts of foods, the study of *ad libitum* food intake measurement was not an option. Finally, in addition to taste, the aroma from the release of volatile compounds from spices could have also contributed to greater appetite suppression. It is well recognized that aroma from foods and non-food materials reaching the olfactory epithelium via the orthonasal and the retronasal routes can lead to appetite suppression [34,35].

## 5. Conclusions

The results from our study indicate that mixed spices consumption can lead to greater increases in inter-meal satiety through suppression in both 'hunger' as well as in 'desire to eat' in the period immediately subsequent to the meal. Both differences in taste intensities as well as aroma from the addition of mixed spices may have contributed to a greater satiety value of the mixed spice containing meals. These changes in postprandial appetite responses seem to be independent of changes in plasma ghrelin concentration, although could potentially have been related to the increase in postprandial plasma GLP-1 concentration, as found within the same cohort previously. The postprandial effects of individual spices or mixed spices on other appetite hormones remain to be established. It should be noted that this was a single-meal, acute feeding study and any long-term effects remain to be further investigated.

**Acknowledgments:** This project was funded by the Singapore Institute for Clinical Sciences, Agency for Science Technology and Research (A*STAR), Singapore. The authors would like to thank Keri McCrickerd for the fruitful discussions and the volunteers for taking part in this study.

**Author Contributions:** S.H. and C.J.H. conceived and designed the experiments; S.C.C. and S.H. performed the experiments; S.H., J.L., and S.P. analyzed the data; S.H., J.L., and S.P. wrote the paper. All authors have read and approved the manuscript.

**Conflicts of Interest:** The authors declare no conflict of interest.

## References

1. Mattes, R.D. Spices and energy balance. *Physiol. Behav.* **2012**, *107*, 584–590. [CrossRef] [PubMed]
2. Warwick, Z.S.; Hall, W.G.; Pappas, T.N.; Schiffman, S.S. Taste and smell sensations enhance the satiating effect of both a high-carbohydrate and a high-fat meal in humans. *Physiol. Behav.* **1993**, *53*, 553–563. [CrossRef]
3. Masic, U.; Yeomans, M.R. Does monosodium glutamate interact with macronutrient composition to influence subsequent appetite? *Physiol. Behav.* **2013**, *116–117*, 23–29. [CrossRef] [PubMed]
4. Sørensen, L.B.; Møller, P.; Flint, A.; Martens, M.; Raben, A. Effect of sensory perception of foods on appetite and food intake: A review of studies on humans. *Int. J. Obes.* **2003**, *27*. [CrossRef] [PubMed]
5. McCrickerd, K.; Forde, C.G. Sensory influences on food intake control: Moving beyond palatability. *Obes. Rev.* **2016**, *17*, 18–29. [CrossRef] [PubMed]
6. Peters, J.C.; Polsky, S.; Stark, R.; Zhaoxing, P.; Hill, J.O. The influence of herbs and spices on overall liking of reduced fat food. *Appetite* **2014**, *79*, 183–188. [CrossRef] [PubMed]
7. Manero, J.; Phillips, C.; Ellison, B.; Lee, S.-Y.; Nickols-Richardson, S.M.; Chapman-Novakofski, K.M. Influence of seasoning on vegetable selection, liking and intent to purchase. *Appetite* **2017**, *116*, 239–245. [CrossRef] [PubMed]
8. Gregersen, N.T.; Belza, A.; Jensen, M.G.; Ritz, C.; Bitz, C.; Hels, O.; Frandsen, E.; Mela, D.J.; Astrup, A. Acute effects of mustard, horseradish, black pepper and ginger on energy expenditure, appetite, ad libitum energy intake and energy balance in human subjects. *Br. J. Nutr.* **2012**, *109*, 556–563. [CrossRef] [PubMed]
9. Ogawa, K.; Ito, M. Appetite-enhancing effects of curry oil. *Biol. Pharm. Bull.* **2016**, *39*, 1559–1563. [CrossRef] [PubMed]
10. Mattes, R.D.; Cowart, B.J. Dietary assessment of patients with chemosensoty disorders. *J. Acad. Nutr. Diet.* **1994**, *94*, 50–56. [CrossRef]
11. Haldar, S.; Chia, S.C.; Lee, S.H.; Lim, J.; Leow, M.K.-S.; Chan, E.C.Y.; Henry, C.J. Polyphenol-rich curry made with mixed spices and vegetables benefits glucose homeostasis in Chinese males (polyspice study): A dose–response randomized controlled crossover trial. *Eur. J. Nutr.* **2017**. [CrossRef] [PubMed]

12. Haldar, S.; Chia, S.C.; Henry, C.J. Polyphenol-rich curry made with mixed spices and vegetables increases postprandial plasma GLP-1 concentration in a dose-dependent manner. *Eur. J. Clin. Nutr.* **2018**, *72*, 297–300. [CrossRef] [PubMed]

13. Gibbons, C.; Caudwell, P.; Finlayson, G.; Webb, D.-L.; Hellström, P.M.; Näslund, E.; Blundell, J.E. Comparison of postprandial profiles of ghrelin, active GLP-1, and total PYY to meals varying in fat and carbohydrate and their association with hunger and the phases of satiety. *J. Clin. Endocrinol. Metab.* **2013**, *98*, E847–E855. [CrossRef] [PubMed]

14. Jakubowicz, D.; Froy, O.; Wainstein, J.; Boaz, M. Meal timing and composition influence ghrelin levels, appetite scores and weight loss maintenance in overweight and obese adults. *Steroids* **2012**, *77*, 323–331. [CrossRef] [PubMed]

15. Gruendel, S.; Garcia, A.L.; Otto, B.; Mueller, C.; Steiniger, J.; Weickert, M.O.; Speth, M.; Katz, N.; Koebnick, C. Carob pulp preparation rich in insoluble dietary fiber and polyphenols enhances lipid oxidation and lowers postprandial acylated ghrelin in humans. *J. Nutr.* **2006**, *136*, 1533–1538. [CrossRef] [PubMed]

16. Panickar, K.S. Effects of dietary polyphenols on neuroregulatory factors and pathways that mediate food intake and energy regulation in obesity. *Mol. Nutr. Food Res.* **2013**, *57*, 34–47. [CrossRef] [PubMed]

17. Kaliora, A.C.; Kanellos, P.T.; Gioxari, A.; Karathanos, V.T. Regulation of gip and ghrelin in healthy subjects fed on sun-dried raisins: A pilot study with a crossover trial design. *J. Med. Food* **2017**, *20*, 301–308. [CrossRef] [PubMed]

18. Serrano, J.; Casanova-Martí, À.; Depoortere, I.; Blay, M.T.; Terra, X.; Pinent, M.; Ardévol, A. Subchronic treatment with grape-seed phenolics inhibits ghrelin production despite a short-term stimulation of ghrelin secretion produced by bitter-sensing flavanols. *Mol. Nutr. Food Res.* **2016**, *60*, 2554–2564. [CrossRef] [PubMed]

19. Hill, A.J.; Magson, L.D.; Blundell, J.E. Hunger and palatability: Tracking ratings of subjective experience before, during and after the consumption of preferred and less preferred food. *Appetite* **1984**, *5*, 361–371. [CrossRef]

20. Flint, A.; Raben, A.; Blundell, J.E.; Astrup, A. Reproducibility, power and validity of visual analogue scales in assessment of appetite sensations in single test meal studies. *Int. J. Obes.* **2000**, *24*. [CrossRef]

21. Chapman, I.M.; Goble, E.A.; Wittert, G.A.; Morley, J.E.; Horowitz, M. Effect of intravenous glucose and euglycemic insulin infusions on short-term appetite and food intake. *Am. J. Physiol.* **1998**, *274*, R596–R603. [CrossRef] [PubMed]

22. Falkén, Y.; Hellström, P.M.; Sanger, G.J.; Dewit, O.; Dukes, G.; Grybäck, P.; Holst, J.J.; Näslund, E. Actions of prolonged ghrelin infusion on gastrointestinal transit and glucose homeostasis in humans. *Neurogastroenterol. Motil.* **2010**, *22*, e192–e200. [CrossRef] [PubMed]

23. Edholm, T.; Degerblad, M.; Grybäck, P.; Hilsted, L.; Holst, J.J.; Jacobsson, H.; Efendic, S.; Schmidt, P.T.; Hellström, P.M. Differential incretin effects of GIP and GLP-1 on gastric emptying, appetite, and insulin-glucose homeostasis. *Neurogastroenterol. Motil.* **2010**, *22*. [CrossRef] [PubMed]

24. Hlebowicz, J.; Hlebowicz, A.; Lindstedt, S.; Björgell, O.; Höglund, P.; Holst, J.J.; Darwiche, G.; Almér, L.-O. Effects of 1 and 3 g cinnamon on gastric emptying, satiety, and postprandial blood glucose, insulin, glucose-dependent insulinotropic polypeptide, glucagon-like peptide 1, and ghrelin concentrations in healthy subjects. *Am. J. Clin. Nutr.* **2009**, *89*, 815–821. [CrossRef] [PubMed]

25. Zanzer, Y.C.; Plaza, M.; Dougkas, A.; Turner, C.; Björck, I.; Östman, E. Polyphenol-rich spice-based beverages modulated postprandial early glycaemia, appetite and pyy after breakfast challenge in healthy subjects: A randomized, single blind, crossover study. *J. Funct. Foods* **2017**, *35*, 574–583. [CrossRef]

26. Maji Amal, K.; Banerji, P. Phytochemistry and gastrointestinal benefits of the medicinal spice, capsicum annuum l (chilli): A review. *J. Complement. Integr. Med.* **2016**, *13*, 97–122. [CrossRef] [PubMed]

27. Ghawi, S.K.; Rowland, I.; Methven, L. Enhancing consumer liking of low salt tomato soup over repeated exposure by herb and spice seasonings. *Appetite* **2014**, *81*, 20–29. [CrossRef] [PubMed]

28. Knaapila, A.; Laaksonen, O.; Virtanen, M.; Yang, B.; Lagström, H.; Sandell, M. Pleasantness, familiarity, and identification of spice odors are interrelated and enhanced by consumption of herbs and food neophilia. *Appetite* **2017**, *109*, 190–200. [CrossRef] [PubMed]

29. Griffioen-Roose, S.; Mars, M.; Finlayson, G.; Blundell, J.E.; de Graaf, C. Satiation due to equally palatable sweet and savory meals does not differ in normal weight young adults. *J. Nutr.* **2009**, *139*, 2093–2098. [CrossRef] [PubMed]

30. Byrnes, N.K.; Hayes, J.E. Behavioral measures of risk tasking, sensation seeking and sensitivity to reward may reflect different motivations for spicy food liking and consumption. *Appetite* **2016**, *103*, 411–422. [CrossRef] [PubMed]
31. Yeomans, M.R. Palatability and the micro-structure of feeding in humans: The appetizer effect. *Appetite* **1996**, *27*, 119–133. [CrossRef] [PubMed]
32. Yeomans, M.R. Taste, palatability and the control of appetite. *Proc. Nutr. Soc.* **1998**, *57*, 609–615. [CrossRef] [PubMed]
33. Bolhuis, D.P.; Lakemond, C.M.M.; de Wijk, R.A.; Luning, P.A.; de Graaf, C. Both longer oral sensory exposure to and higher intensity of saltiness decrease ad libitum food intake in healthy normal-weight men. *J. Nutr.* **2011**, *141*, 2242–2248. [CrossRef] [PubMed]
34. Ramaekers, M.G.; Luning, P.A.; Ruijschop, R.M.A.J.; Lakemond, C.M.M.; Bult, J.H.F.; Gort, G.; van Boekel, M.A.J.S. Aroma exposure time and aroma concentration in relation to satiation. *Br. J. Nutr.* **2013**, *111*, 554–562. [CrossRef] [PubMed]
35. Ruijschop, R.M.A.J.; Boelrijk, A.E.M.; de Graaf, C.; Westerterp-Plantenga, M.S. Retronasal aroma release and satiation: A review. *J. Agric. Food Chem.* **2009**, *57*, 9888–9894. [CrossRef] [PubMed]

# foods

MDPI

*Article*

# Lactic Acid Bacteria from Kefir Increase Cytotoxicity of Natural Killer Cells to Tumor Cells

**Takuya Yamane [1,2,3,*], Tatsuji Sakamoto [1,2], Takenori Nakagaki [3] and Yoshihisa Nakano [1,2]**

[1]   Center for Research and Development Bioresources, Organization for Research Promotion,
     Osaka Prefecture University, Sakai, Osaka 599-8570, Japan; sakamoto@biochem.osakafu-u.ac.jp (T.S.);
     nakano@biochem.osakafu-u.ac.jp (Y.N.)
[2]   Department of Applied Life Sciences, Graduate School of Life and Environmental Sciences,
     Osaka Prefecture University, Sakai, Osaka 599-8531, Japan
[3]   Institute of Food Sciences, Nakagaki Consulting Engineer and Co., Ltd., Nishi-ku, Sakai 593-8328, Japan;
     tnakagaki@nakagaki.co.jp
*   Correspondence: tyt29194@osakafu-u.ac.jp; Tel.: +81-72-254-9937 (ext. 2962); Fax: +81-72-254-9937

Received: 26 February 2018; Accepted: 25 March 2018; Published: 27 March 2018

**Abstract:** The Japanese fermented beverage, homemade kefir, contains six lactic acid bacteria: *Lactococcus. lactis* subsp. *Lactis*, *Lactococcus. lactis* subsp. *Cremoris*, *Lactococcus. Lactis* subsp. *Lactis biovar diacetylactis*, *Lactobacillus plantarum*, *Leuconostoc meseuteroides* subsp. *Cremoris* and *Lactobacillus casei*. In this study, we found that a mixture of the six lactic acid bacteria from kefir increased the cytotoxicity of human natural killer KHYG-1 cells to human chronic myelogenous leukemia K562 cells and colorectal tumor HCT116 cells. Furthermore, levels of mRNA expression and secretion of IFN-$\gamma$ (interferon gamma) increased in KHYG-1 cells that had been treated with the six lactic acid bacteria mixture from kefir. The results suggest that the six lactic acid bacteria mixture from kefir has strong effects on natural immunity and tumor cell cytotoxicity.

**Keywords:** kefir; lactic acid bacteria; natural killer cell; cytotoxicity

## 1. Introduction

The fermented beverage kefir has many health benefits, including its antibacterial, anticarcinogenic, wound healing, antiallergenic, immunomodulation, and gastrointestinal immunity effects [1]. The fermented beverage, homemade kefir, has been available for more than 20 years in Japan. Homemade kefir contains six lactic acid bacteria: *Lactococcus. lactis* subsp. *Lactis* [2–12], *Lactococcus. lactis* subsp. *Cremoris* [8,9,13], *Lactococcus. Lactis* subsp. *Lactis biovar diacetylactis* [4], *Lacto bacillus planterun* [4,14–16], *Leuconostoc meseuteroides* subsp. *Cremoris* [7,9], and *Lacto bacillus casei* [2,5,12,16]. Kefir is able to offer health benefits through modulation of the gastrointestinal immune system, and previous studies showed that kefir increases the levels of some proinflammatory cytokines including TNF-$\alpha$ (tumor necrosis factor alpha), IFN-$\gamma$ (interferon gamma) and IL-12 (interleukin 12) [1]. Lactic acid bacteria have beneficial effects with respect to diarrhea [17,18], food allergies [19], and inflammatory bowel disease [20–22]. Lactic acid bacteria also play an important role in the prevention of colorectal cancer [23,24], and *Lacto bacillus casei* Shirota increases the activity of natural killer (NK) cells [25,26]. However, the effect of homemade kefir on NK cell activity is not clear. In this study, we found that a six lactic acid bacteria mixture from homemade kefir increases NK cell activity, including cytotoxic activity against human colorectal tumor HCT116 cells. These results suggest that the mixture of six lactic acid bacteria from homemade kefir has important effects on innate immunity and tumor cell cytotoxicity.

## 2. Materials and Methods

### 2.1. Materials

Homemade kefir was provided by Nakagaki Consulting Engineer and Co., Ltd. (Osaka, Japan). Roswell Park Memorial Institute (RPMI) 1640 medium, Dulbecco's modified Eagle's medium (DMEM), and human interleukin-2 were purchased from Wako Pure Chemical Industries, Ltd. (Osaka, Japan). All other reagents were of analytical grade.

### 2.2. Cell Culture

KHYG-1 and K562 cells were provided by JCRB Cell Bank (Osaka, Japan), and HCT116 cells were purchased from American Tissue Culture Collection (ATCC) (Manassas, VA, USA). KHYG-1 cells were cultured in RPMI 1640 medium with 10% fetal bovine serum (Sigma, St. Louis, MO, USA) and 1 μg/mL IL-2. K562 and HCT116 cells were cultured in RPMI 1640 medium and in Dulbecco's modified Eagle's medium (DMEM), respectively, with 10% fetal bovine serum.

### 2.3. Preparation of Lactic Acid Bacteria Mixture

One gram of homemade kefir powder was added to 1 L of de Man, Rogosa and Sharpe (MRS) broth and cultured for 24 h at 22.5 °C. The cultured solution was centrifuged at 3000 rpm for 30 min at 4 °C using a TOMY centrifugation apparatus (TOMY SEIKO, Tokyo, Japan), and the six lactic acid bacteria mixture was obtained from the precipitates. A lactic acid bacteria mixture from killed bacteria was prepared by autoclaving at 120 °C for 15 min. Identification of the lactic acid bacteria cultured in the MRS broth was performed by polymerase chain reaction (PCR) using DNA from the six lactic acid bacteria as templates. After PCR, amplified DNA was separated on 2.0% agarose gels and stained with ethidium bromide. The PCR conditions and primers used are shown in Table 1.

### 2.4. Measurement of NK Cell Activity and Tumor Cytotoxicity

KHYG-1 cells were treated with the six lactic acid bacteria mixture for 24 h at 37 °C in a 5% $CO_2$ incubator. The cells that had been treated with the six lactic acid bacteria mixture were reacted with K562 cells or with HCT116 cells for 4 h at 37 °C in a 5% $CO_2$ incubator. NK cell activity and tumor cytotoxicity were measured using an LDH Cytotoxicity Detection Kit (Takara, Shiga, Japan).

### 2.5. RT-PCR (Reverse Transcription Polymerase Chain Reaction)

Nucleotide sequences of primers used for RT-PCR were as follows: hIFNγ-F: 5′-GAATGTCCAACGCAAAGCAA-3′, hIFNγ-R: 5′-GCTGCTGGCGACAGTTCAG-3′, hACTB-F: 5′-CTTCCTGGGCATGGAGTC-3′, and hACTB-R: 5′-GGATGTCCACGTCACACTTC-3′. The full RNAs were prepared from cells and subjected to semi-quantitative RT-PCR analyses. PCR conditions were 1 min at 94 °C, 30 s at 94 °C, 30 s at 60 °C, and 35 cycles of 1 min at 72 °C for IFNγ and ACTB. After the reactions, PCR products were separated on 2.0% agarose gels and stained with SAFELOOK™ Pregreen (Wako, Osaka, Japan). Intensities of the bands were quantified using Image J software (https://imagej.nih.gov/ij/). β-actin mRNA was also amplified as a control.

### 2.6. IFN-γ Measurement

IFN-γ was measured using a Human IFN-γ ELISA (enzyme-linked immunosorbent assay) Development Kit (PeproTech, Rocky Hill, NJ, USA) according to the manufacturer's instructions.

### 2.7. Statistical Analysis

Data are expressed as means ± S.E (standard error). Statistical analyses were performed with an analysis of variance (one-way ANOVA) followed by an unpaired Student's *t*-test.

**Table 1.** Polymerase chain reaction (PCR) primers and reacting conditions.

| Strain | Primer Sequence | | PCR Condition | Ref. |
|---|---|---|---|---|
| L. lactis subsp. Lactis | sense | 5'-CGTTATGGGATTTGATGGGATATAAAGC-3' | 94 °C 9 min, 94 °C 30 s, 50 °C 30 s, 72 °C 60 s × 45 cycles, 72 °C 7 min | [27] |
| | antisense | 5'-ACTCTTCTTAAGAACAAGTTTAACAGC-3' | | |
| L. lactis subsp. Cremoris | sense | 5'-CGTTATGGGATTTGATGGGATATAAAGC-3' | 94 °C 9 min, 94 °C 30 s, 50 °C 30 s, 72 °C 60 s × 45 cycles, 72 °C 7 min | |
| | antisense | 5'-ACTCTTCTTAAGAACAAGTTTAACAGC-3' | | |
| L. lactis subsp. lactis biovar diacetylactis | sense | 5'-CTTCGTTATGATTTTACA-3' | 94 °C 30 s, 94 °C 30 s, 46 °C 60 s, 72 °C 60 s × 30 cycles, 72 °C 2 min | [28,29] |
| | antisense | 5'-AATATCAACAATTCCATG-3' | | |
| Leuconostoc mesenteroides subsp. Cremoris (Genus-specific primers for Leuconostoc) | sense | 5'-AACTTAGTGTCGCATGAC-3' | 94 °C 5 min, 94 °C 60 s, 60 °C 60 s, 72 °C 60 s × 30 cycles, 72 °C 10 min | [30] |
| | antisense | 5'-AGTCGAGTTACAGACTACAA-3' | | |
| Lactobacillus plantarum | sense | 5'-CCGTTTATGCGGAACACCTA-3' | 94 °C 3 min, 94 °C 30 s, 54 °C 10 s, 72 °C 30 s × 30 cycles, 72 °C 5 min | [31] |
| | antisense | 5'-TCGGGATTACCAAACATCAC-3' | | |
| Lactobacillus casei | sense | 5-TGCACTGGAGATTCGACTTAA-3 | 94 °C 3 min, 94 °C 45 s, 45 °C 45 s, 72 °C 60 s × 30 cycles, 72 °C 5 min | [32] |

## 3. Results and Discussion

### 3.1. Induction of NK Activity by the Six Lactic Acid Bacteria from Homemade-Kefir

As shown in Figure 1, the six lactic acid bacteria from homemade kefir were detected in MRS broth cultures and identified using PCR with the bacteria's DNA as templates. The cytotoxicity of KHYG-1 cells that had been treated with the six lactic acid bacteria mixture, against K562 cells, increased in a cell number-dependent manner (Figure 2). As shown in Figure 3A, KHYG-1 cells were activated by the six lactic acid bacteria mixture and increased their cytotoxicity against K562 cells in a dose-dependent manner. Lactic acid bacteria have many effects on biological activities such as apoptosis induction, antioxidant activities, and immune response improvement [33]. Nagao et al. reported that *Lactobacillus casei Shirota* enhanced NK cell activity [25]. *Lacto bacillus plantarum* C4 isolated from kefir has also been shown to prevent infection with *Yersinia enterocolitice* O9 in the intestine in mice. Additionally, a concentrated supernatant from the cultured medium of *Lacto bacillus plantarum* C4 has antibacterial and anti-tumor activities [34]. Since the six lactic acid bacteria mixture from homemade kefir includes *Lacto bacillus plantarum* and *Lacto bacillus casei*, it is thought that these lactic acid bacteria contribute to the induction of NK cell activity.

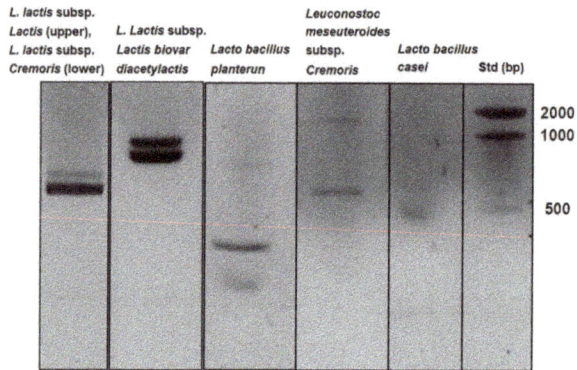

**Figure 1.** Identification of the six lactic acid bacteria cultured in de Man, Rogosa and Sharpe (MRS) broth using polymerase chain reaction (PCR) as described in the Materials and Methods. The amplified DNA after PCR were separated on agarose gels.

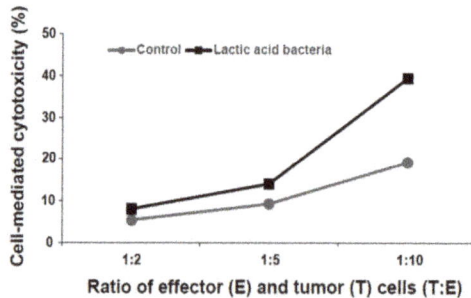

**Figure 2.** Cell-mediated cytotoxicity of natural killer (NK) cells by a six lactic acid bacteria mixture. KHYG-1 cells were treated with a six lactic acid bacteria mixture for 24 h at 37 °C in a 5% $CO_2$ incubator. The treated cells were reacted with K562 cells for 4 h at 37 °C in a 5% $CO_2$ incubator. Cell-mediated cytotoxicity increased for KHYG-1 cells that had been treated with the six lactic acid bacteria mixture.

**A**

**B**

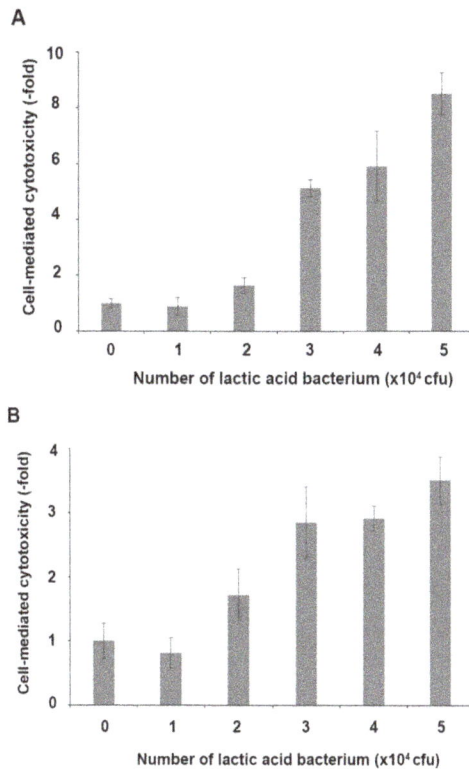

**Figure 3.** Induction of cell-mediated cytotoxicity by a six lactic acid bacteria mixture. KHYG-1 cells were treated with a six lactic acid bacteria mixture for 24 h at 37 °C in a 5% $CO_2$ incubator. The treated cells were reacted with K562 cells (**A**) or with HCT116 cells (**B**) for 4 h at 37 °C in a 5% $CO_2$ incubator. In both cases, cell-mediated cytotoxicity was increased by KHYG-1 cells that had been treated with the six lactic acid bacteria mixture in a dose-dependent manner.

*3.2. Induction of Cytotoxicity by the Six Lactic Acid Bacteria Against Human Colorectal Tumor Cells*

Cytotoxic activity of KHYG-1 cells against HCT116 cells was increased by the six lactic acid bacteria mixture in a dose-dependent manner (Figure 3B). NK cells have the intrinsic ability to detect and kill cancer cells [35]. The lytic functions of NK cells mediated by granzyme B and perforin have dominated investigations of the role of NK cells in anti-cancer defense [36]. The effector function of NK cells is tightly regulated by activating and inhibiting receptors. Tumor cells can evade the anti-tumor activity of NK cells by modifying the expression of NK cell receptors. In colorectal carcinoma, for instance, the expression of activating receptors—such as CD16 (FCGR3A, Fc fragment of IgG receptor IIIa), NKp46 (NCR1, natural cytotoxicity triggering receptor 1), NKp30 (NCR3, natural cytotoxicity triggering receptor 3), and NKG2D (KLRK1, killer cell lectin like receptor K1)—is decreased. Additionally, the expression of inhibiting receptors, such as KIR3DL1 (killer cell immunoglobulin-like receptor 3DL1) and NKG2A (killer cell lectin like receptor C1), is increased [37]. The six lactic acid bacteria mixture or its secretory compounds may regulate the expression of these receptor genes and inhibit tumor formation by escaping from NK cell-mediated cytotoxicity. Lactic acid bacteria have indeed been demonstrated to have a host of properties for preventing the development of a colorectal tumor, specifically by inhibiting its initiation or progression. In the pathway of the anticancer immune response, lactic acid bacteria also stimulate NK cell activity [33]. Furthermore, a seminal 11-year prospective cohort study conducted in Japan showed an inverse correlation between increase of cancer incidence and the level of NK cell cytotoxicity [38]. These results indicate that the six lactic acid bacteria mixture from homemade kefir reacts with human colorectal tumor HCT116 cells through activation of the immunity of NK cells.

### 3.3. Induction of Cytotoxicity of KHYG-1 Cells by the Killed Lactic Acid Bacteria Mixture

To examine the distinct effects of live and dead lactic acid bacteria on NK cell activation, KHYG-1 cells were treated with the live and the killed six lactic acid bacteria mixtures and cytotoxicity assays were carried out. As shown in Figure 4A, activation of KHYG-1 cells was increased by treatment with both the live and killed mixtures. Cytotoxicity of KHYG-1 cells against HCT116 cells was also increased by both mixtures (Figure 4B). Previous studies, on the other hand, have shown that 40% to 62% of several strains of *L. plantarum* survive under the low pH conditions in a simulated intestine [39], and that 81% to 93% of several strains of *L. casei* can also survive at pH 2.0 for 2 h [40]. The present and previous results, therefore, suggest that both lactic acid bacteria that have been killed by gastric acid and live lactic acid bacteria have effects on NK cell activation in the intestine after the administration of homemade kefir.

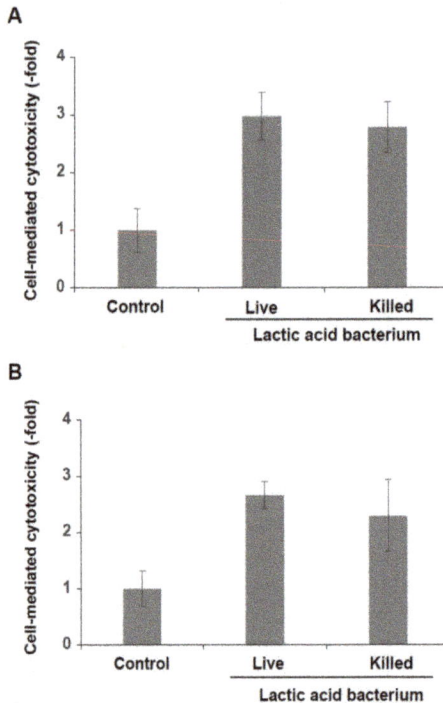

**Figure 4.** Induction of cell-mediated cytotoxicity of KHYG-1 cells by the live and the killed lactic acid bacteria mixtures. KHYG-1 cells were treated with both mixtures ($3 \times 10^4$ cfu) for 24 h at 37 °C in a 5% $CO_2$ incubator. The treated cells were reacted with K562 cells (**A**) or with HCT116 cells (**B**) for 4 h at 37 °C in a 5% $CO_2$ incubator. In both cases, cell-mediated cytotoxicity was increased in KHYG-1 cells that had been treated with both the live and the killed lactic acid bacteria mixtures.

## 3.4. Increased Expression and Secretion Levels of IFN-γ in KHYG-1 Cells

IFN-γ mRNA and IFN-γ secretion were both increased in KHYG-1 cells that had been treated by the six lactic acid bacteria mixture (Figure 5A,B). NK cells regulate various immune cells through the secretion of IFN-γ and they perform tumor surveillance [41]. Increased IFN-γ expression and secretion induced by the mixture of six lactic acid bacteria from homemade kefir may provide a synergistic effect on anti-tumor activity.

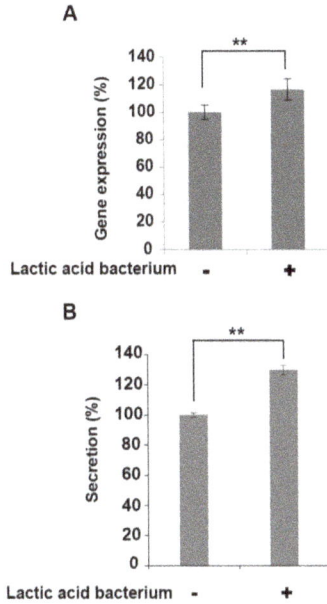

**A**

**B**

**Figure 5.** Increased expression and secretion levels of IFN-γ. (**A**) The expression level of IFN-γ mRNA was increased in KHYG-1 cells that had been treated with the six lactic acid bacteria mixture ($3 \times 10^4$ cfu) for 4 h at 37 °C in a 5% $CO_2$ incubator ($n = 3$, ** $p < 0.01$). (**B**) The secretion level of IFN-γ mRNA was increased in the same, treated, KHYG-1 cells ($n = 6$, ** $p < 0.01$).

## 4. Conclusions

This study showed that the cytotoxicity of NK cells and the expression and secretion of IFN-γ in NK cells were increased after treatment of the cells with a six lactic acid bacteria mixture. Further in vitro and in vivo studies using animal models of colorectal cancer are needed to investigate the mechanisms underlying the effects of the six lactic acid bacteria mixture.

**Author Contributions:** Takuya Yamane, Tatsuji Sakamoto, Takenori Nakagaki and Yoshihisa Nakano conceptualized the study; Takuya Yamane carried out experiments; Takuya Yamane conducted the analyses; and Takuya Yamane wrote the manuscript.

**Conflicts of Interest:** The authors declare no conflicts of interest.

## References

1. Bourrie, B.C.; Willing, B.P.; Cotter, P.D. The Microbiota and Health Promoting Characteristics of the Fermented Beverage Kefir. *Front. Microbiol.* **2016**, *7*, 647. [CrossRef] [PubMed]
2. Angulo, L.; Lopez, E.; Lema, C. Microflora present in kefir grains of the Galician region (North-West of Spain). *J. Dairy Res.* **1993**, *60*, 263–267. [CrossRef] [PubMed]

3. Pintado, M.E.; Da Silva, J.A.L.; Fernandes, P.B.; Malcata, F.X.; Hogg, T.A. Microbiological and rheological studies on Portuguese kefir grains. *Int. J. Food Sci. Technol.* **1996**, *31*, 15–26. [CrossRef]
4. Garrote, G.L.; Abraham, A.G.; De Antoni, G.L. Chemical and microbiological characterisation of kefir grains. *J. Dairy Res.* **2001**, *68*, 639–652. [CrossRef] [PubMed]
5. Simova, E.; Beshkova, D.; Angelov, A.; Hristozova, T.; Frengova, G.; Spasov, Z. Lactic acid bacteria and yeasts in kefir grains and kefir made from them. *J. Ind. Microbiol. Biotechnol.* **2002**, *28*, 1–6. [CrossRef] [PubMed]
6. Witthuhn, R.C.; Schoeman, T.; Britz, T.J. Isolation and characterization of the microbial population of different South African kefir grains. *Int. J. Dairy Technol.* **2004**, *57*, 33–37. [CrossRef]
7. Witthuhn, R.; Schoeman, T.; Britz, T. Characterisation of the microbial population at different stages of Kefir production and Kefir grain mass cultivation. *Int. Dairy J.* **2005**, *15*, 383–389. [CrossRef]
8. Yüksekdağ, Z.; Beyatli, Y.; Aslim, B. Determination of some characteristics coccoid forms of lactic acid bacteria isolated from Turkish kefirs with natural probiotic. *LWT Food Sci. Technol.* **2004**, *37*, 663–667. [CrossRef]
9. Mainville, I.; Robert, N.; Lee, B.; Farnworth, E.R. Polyphasic characterization of the lactic acid bacteria in kefir. *Syst. Appl. Microbiol.* **2006**, *29*, 59–68. [CrossRef] [PubMed]
10. Chen, H.-C.; Wang, S.-Y.; Chen, M.-J. Microbiological study of lactic acid bacteria in kefir grains by culture-dependent and culture-independent methods. *Food Microbiol.* **2008**, *25*, 492–501. [CrossRef] [PubMed]
11. Garofalo, C.; Osimani, A.; Milanović, V.; Aquilanti, L.; De Filippis, F.; Stellato, G.; Di Mauro, S.; Turchetti, B.; Buzzini, P.; Ercolini, D.; et al. Bacteria and yeast microbiota in milk kefir grains from different Italian regions. *Food Microbiol.* **2015**, *49*, 123–133. [CrossRef] [PubMed]
12. Zanirati, D.F.; Abatemarco, M.; de Cicco Sandes, S.H.; Nicoli, J.R.; Nunes, Á.C.; Neumann, E. Selection of lactic acid bacteria from Brazilian kefir grains for potential use as starter or probiotic cultures. *Anaerobe* **2015**, *32*, 70–76. [CrossRef] [PubMed]
13. Korsak, N.; Taminiau, B.; Leclercq, M.; Nezer, C.; Crevecoeur, S.; Ferauche, C.; Detry, E.; Delcenserie, V.; Daube, G. Short communication: Evaluation of the microbiota of kefir samples using metagenetic analysis targeting the 16S and 26S ribosomal DNA fragments. *J. Dairy Sci.* **2015**, *98*, 3684–3689. [CrossRef] [PubMed]
14. Santos, A.; San Mauro, M.; Sanchez, A.; Torres, J.M.; Marquina, D. The antimicrobial properties of different strains of *Lactobacillus* spp. isolated from kefir. *Syst. Appl. Microbiol.* **2003**, *26*, 434–437. [CrossRef] [PubMed]
15. Da Cruz Pedrozo Miguel, M.G.; Cardoso, P.G.; de Assis Lago, L.; Schwan, R.F. Diversity of bacteria present in milk kefir grains using culture-dependent and culture-independent methods. *Food Res. Int.* **2010**, *43*, 1523–1528. [CrossRef]
16. Nalbantoglu, U.; Cakar, A.; Dogan, H.; Abaci, N.; Ustek, D.; Sayood, K.; Can, H. Metagenomic analysis of the microbial community in kefir grains. *Food Microbiol.* **2014**, *41*, 42–51. [CrossRef] [PubMed]
17. Chouraqui, J.P.; Van Egroo, L.D.; Fichot, M.C. Acidified milk formula supplemented with *Bifidobacterium lactis*: Impact on infant diarrhea in residential care settings. *J. Pediatr. Gastroenterol. Nutr.* **2004**, *38*, 288–292. [CrossRef] [PubMed]
18. Gaón, D.; García, H.; Winter, L.; Rodríguez, N.; Quintás, R.; González, S.N.; Oliver, G. Effect of *Lactobacillus* strains and *Saccharomyces boulardii* on persistent diarrhea in children. *Medicina (Buenos Aires)* **2003**, *63*, 293–298.
19. Pohjavuori, E.; Viljanen, M.; Korpela, R.; Kuitunen, M.; Tiittanen, M.; Vaarala, O.; Savilahti, E. *Lactobacillus* GG effect in increasing IFN-gamma production in infants with cow's milk allergy. *J. Allergy Clin. Immunol.* **2004**, *114*, 131–136. [CrossRef] [PubMed]
20. Bourlioux, P.; Koletzko, B.; Guarner, F.; Braesco, V. The intestine and its microflora are partners for the protection of the host: Report on the Danone Symposium "The Intelligent Intestine," held in Paris, June 14, 2002. *Am. J. Clin. Nutr.* **2003**, *78*, 675–683. [CrossRef] [PubMed]
21. Azcárate-Peril, M.A.; Sikes, M.; Bruno-Bárcena, J.M. The intestinal microbiota, gastrointestinal environment and colorectal cancer: A putative role for probiotics in prevention of colorectal cancer? *Am. J. Physiol. Gastrointest. Liver Physiol.* **2011**, *301*, G401–G424. [CrossRef] [PubMed]
22. Del Carmen, S.; de LeBlanc, A.M.; Miyoshi, A.; Rocha, C.S.; Azevedo, V.; LeBlanc, J.G. Potential application of probiotics in the prevention and treatment of inflammatory bowel diseases. *Ulcers* **2011**, *2011*, 841651. [CrossRef]
23. Rafter, J.J. The role of lactic acid bacteria in colon cancer prevention. *Scand. J. Gastroenterol.* **1995**, *30*, 497–502. [CrossRef] [PubMed]

24. Hirayama, K.; Rafter, J. The role of lactic acid bacteria in colon cancer prevention: Mechanistic considerations. *Antonie Van Leeuwenhoek* **1999**, *76*, 391–394. [PubMed]
25. Nagao, F.; Nakayama, M.; Muto, T.; Okumura, K. Effects of a fermented milk drink containing *Lactobacillus casei* strain Shirota on the immune system in healthy human subjects. *Biosci. Biotechnol. Biochem.* **2000**, *64*, 2706–2708. [CrossRef] [PubMed]
26. Dong, H.; Rowland, I.; Tuohy, K.M.; Thomas, L.V.; Yaqoob, P. Selective effects of *Lactobacillus casei* Shirota on T cell activation, natural killer cell activity and cytokine production. *Clin. Exp. Immunol.* **2010**, *161*, 378–388. [CrossRef] [PubMed]
27. Nomura, M.; Kobayashi, M.; Okamoto, T. Rapid PCR-Based Method Which Can Determine Both Phenotype and Genotype of *Lactococcus lactis* Subspecies. *Appl. Environ. Microbiol.* **2002**, *68*, 2209–2213. [CrossRef] [PubMed]
28. Beimfohr, C.; Ludwig, W.; Schleifer, K.H. Rapid Genotypic Differentiation of *Lactococcus lactis* Subspecies and Biovar. *Syst. Appl. Microbiol.* **1997**, *20*, 216–221. [CrossRef]
29. Feutry, F.; Torre, P.; Arana, I.; Garcia, S.; Desmasures, N.; Casalta, E. *Lactococcus lactis* strains from raw ewe's milk samples from the PDO Ossau-Iraty cheese area: Levels, genotypic and technological diversity. *Dairy Sci. Technol.* **2012**, *92*, 655–670. [CrossRef]
30. Kahala, M.; Maki, M.; Lehtovaara, A.; Tapanainen, J.M.; Katiska, R.; Juuruskorpi, M.; Juhola, J.; Joutsjoki, V. Characterization of starter lactic acid bacteria from the Finnish fermented milk product viili. *J. Appl. Microbiol.* **2008**, *105*, 1929–1938. [CrossRef] [PubMed]
31. Torriani, S.; Felis, G.E.; Dellaglio, F. Differentiation of *Lactobacillus plantarum*, *L. pentosus*, and *L. paraplantarum* by *recA* gene sequence analysis and multiplex PCR assay with *recA* gene-derived primers. *Appl. Environ. Microbiol.* **2001**, *67*, 3450–3454. [CrossRef] [PubMed]
32. Ward, L.J.H.; Timmins, M.J. Differentiation of *Lactobacillus casei*, *Lactobacillus paracasei* and *Lactobacillus rhamnosus* by polymerase chain reaction. *Lett. Appl. Microbiol.* **1999**, *29*, 90–92. [CrossRef] [PubMed]
33. Zhong, L.; Zhang, X.; Covasa, M. Emerging roles of lactic acid bacteria in protection against colorectal cancer. *World J. Gastroenterol.* **2014**, *20*, 7878–7886. [CrossRef] [PubMed]
34. De Montijo-Prieto, S.; Moreno, E.; Bergillos-Meca, T.; Lasserrot, A.; Ruiz-López, M.D.; Ruiz-Bravo, A.; Jiménez-Valera, M.A. *Lactobacillus plantarum* strain isolated from kefir protects against intestinal infection with *Yersinia enterocolitica* O9 and modulates immunity in mice. *Res. Microbiol.* **2015**, *166*, 626–632. [CrossRef] [PubMed]
35. Nicholson, S.E.; Keating, N.; Belz, G.T. Natural killer cells and anti-tumor immunity. *Mol. Immunol.* **2017**, in press. [CrossRef] [PubMed]
36. Voskoboinik, I.; Whisstock, J.C.; Trapani, J.A. Perforin and granzymes: Function, dysfunction and human pathology. *Nat. Rev. Immunol.* **2015**, *15*, 388–400. [CrossRef] [PubMed]
37. Lee, H.H.; Kang, H.; Cho, H. Natural killer cells and tumor metastasis. *Arch. Pharm. Res.* **2017**, *40*, 1037–1049. [CrossRef] [PubMed]
38. Imai, K.; Matsuyama, S.; Miyake, S.; Suga, K.; Nakachi, K. Natural cytotoxic activity of peripheral-blood lymphocytes and cancer incidence: An 11-year follow-up study of a general population. *Lancet* **2000**, *356*, 1795–1799. [CrossRef]
39. Tokatlı, M.; Gülgör, G.; Bağder Elmacı, S.; Arslankoz İşleyen, N.; Özçelik, F. In Vitro Properties of Potential Probiotic Indigenous Lactic Acid Bacteria Originating from Traditional Pickles. *BioMed Res. Int.* **2015**, *2015*, 315819. [CrossRef] [PubMed]
40. Zhang, B.; Wang, Y.; Tan, Z.; Li, Z.; Jiao, Z.; Huang, Q. Screening of Probiotic Activities of Lactobacilli Strains Isolated from Traditional Tibetan Qula, A Raw Yak Milk Cheese. *Asian-Australas. J. Anim. Sci.* **2016**, *29*, 1490–1499. [CrossRef] [PubMed]
41. Ikeda, H.; Old, L.J.; Schreiber, R.D. The roles of IFN gamma in protection against tumor development and cancer immunoediting. *Cytokine Growth Factor Rev.* **2002**, *13*, 95–109.

foods

MDPI

*Review*

# Betaine in Cereal Grains and Grain-Based Products

**Bojana Filipčev \*, Jovana Kojić, Jelena Krulj, Marija Bodroža-Solarov and Nebojša Ilić**

Institute of Food Technology, University of Novi Sad, 21000 Novi Sad, Serbia; jovana.kojic@fins.uns.ac.rs (J.Ko.); jelena.krulj@fins.uns.ac.rs (J.Kr.); marija.bodroza@fins.uns.ac.rs (M.B.-S.); nebojsa.ilic@fins.uns.ac.rs (N.I.)
\* Correspondence: bojana.filipcev@fins.uns.ac.rs; Tel.: +381-21-485-3778

Received: 27 February 2018; Accepted: 23 March 2018; Published: 29 March 2018

**Abstract:** Betaine is a non-essential nutrient which performs several important physiological functions in organisms. Abundant data exist to suggest that betaine has a potential for prevention of chronic diseases and that its dietary intake may contribute to overall health enhancement. Several studies have pointed out that the betaine status of the general population is inadequate and have suggested nutritional strategies to improve dietary intake of betaine. Cereal-based food has been implicated as the major source of betaine in the Western diet. This review summarizes the results on the betaine content in various cereals and related products. Attention has been given to the betaine content in gluten-free grains and products. It also discusses the stability of betaine during processing (cooking, baking, extrusion) and possibilities to increase betaine content by fortification.

**Keywords:** betaine; cereals; pseudocereals; gluten-free; stability; cooking; baking; extrusion

---

## 1. Introduction

Betaine (*N,N,N*-trimethylglycine, glycine betaine) is an organic nitrogenous compound, found for the first time in sugar beet juice (*Beta vulgaris*).

Betaine is a zwitterion of quaternary ammonium which is still named trimethylglycine and glycine betaine (Figure 1). It is a methyl derivative of the amino acid glycine (($CH_3$)$_3$N$^+$CH$_2$COO$^-$ and molecular weight 117.2). It is characterized as methylamine due to its three free methyl groups [1].

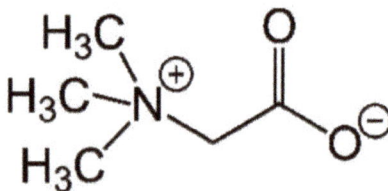

**Figure 1.** Betaine chemical structure.

Various analogues of glycine betaine exist in plants: proline betaine (stachydrine), trigonelline, arsenobetaine, betonicine, butirobetaine, ergothionine, propionobetaine, and sulfur analogues. The sulfur analogues are several in type: β-alaninebetaine, dimethylsulfonioacetate, and dimethylsulfoniopropionate (DMSP). The food survey study by de Zwart et al. [2] showed that only some betaine analogues were present in food at appreciable levels (>10 µg/g)—glycine betaine, proline betaine, trigonelline, and DMSP. Slow et al. [3] indicated glycine betaine as dominant in grain products, proline betaine in citruses, and trigonelline in coffee. Most recently, some rare forms of betaine were identified in the grains of most common cereals: pipercolic acid betaine in rye flour and valine betaine and glutamine betaine in flours of barley, rye, oat, durum, and winter wheat [4].

The content of betaine analogues was found to be vastly variable in grains; higher betaine levels seem to be induced by plant growth under stress conditions (drought, salt stress, cold, freezing, hypoxia, etc.) [2,3]. Since the potential health effects of betaine analogues, particularly trigonelline and proline, have not yet been fully resolved, currently only glycine betaine has dietary relevance.

Betaine represents a bioactive compound that has significant physiological functions in the human organism as an osmolite and donor of methyl groups for many biochemical processes. As such, it is indispensable to preserve the health of kidneys, liver, and heart [5]. This compound has an important role in preventing and treating many chronic diseases, among which lowering of plasma homocysteine levels has gained the most attention [5–7]. High serum homocysteine levels have been associated with increased risk for cardiovascular diseases (stroke, heart attack, atherosclerosis), cancer, peripheral neuropathy, etc. Moreover, betaine has been shown to improve athletic performance by enhancing muscle endurance [7,8].

Although betaine is not an essential nutrient, it cannot be synthesized in adequate quantities by the human body. Humans may obtain betaine from foods rich in betaine or choline or by oral supplements contained with pure preparations. The main sources of betaine in human nutrition are wheat bran, wheat germ, and spinach [9,10]. Several studies denoted cereal foods as major contributors of betaine [3,11–13]. The betaine intake from foods was estimated in a few food surveys (overviewed by Ross et al. [11]) which differ in methodologies used to collect food consumption data, and used different food databases to calculate the betaine intake. Wide gender, national, and international variations were observed [11]. The overall mean betaine intake estimated from these surveys was 131 mg/day [11]. Daily supply of betaine should reach 1500 mg [14] for the manifestation of health effects, so, obviously, the dietary betaine intakes of general populations can be considered to be on the low side. Elderly population (aged over 50 years) or other vulnerable groups (diabetic and homocystinuria patients) may be at a higher risk of inadequate intake. Therefore, many nutritionists advocate for betaine supplementation. Research by Olthof et al. [6] inferred that betaine delivery via food and oral administration of betaine supplement has a similar health effect, where a diet rich in betaine (betaine intake of about 2000 mg/day) had a higher effect on lowering homocysteine than a betaine poor diet (500 mg/day of betaine). It was reported that a betaine-rich meal providing $\approx$800 mg/day betaine exerted similar acute health effects (increased circulating betaine concentrations and lessened post-methionine load rise in homocysteine) as did the $\approx$1 g/day supplement [15]. This supports the idea of dietary adjustments to improve the betaine status of general population. These adjustments may span from an effort to include betaine-rich ingredients in the daily individual diet or to imply strategies of food fortification with betaine.

In the US, betaine is recognized as the Generally Recognized as Safe (GRAS) ingredient while in Europe it has been approved by the European Commission for use in food. In 2012, the European Union Regulatory Authority (Commission Regulation (EU) No. 432/2012) [14] permitted the declaration of a health claim on foods containing at least 500 mg betaine per serving, indicating that health effects may be expected if 1500 mg of betaine is administered daily. The medical statement reads: "Betaine contributes to the normal metabolism of homocysteine". However, this claim should be accompanied with a restriction due to risks associated with excessive intake of betaine: "In order to bear the claim information shall be given to the consumer that a daily intake in excess of 4 g may significantly increase blood cholesterol levels" [14].This work aims to summarize the current findings on the levels of betaine found in cereals and pseudocereals as well as in related products.

## 2. Experimental Methods used in the Analysis of Betaine

In order to determine the content of betaine in food, different methods have been developed. The most common is the liquid chromatography method, but there is no universal method that can be applied to all food matrices. Saarinen et al. [16] analyzed the content of betaine in the chicken liver using a cation exchange column $Ca^{2+}$ and a refractometric detector, although quantification was limited due to poor sensitivity of the detector. Considering physical-chemical properties of betaine, it cannot be analyzed by conventional reversed phase liquid chromatography. Also, betaine has less absorption in the UV–visible spectrum and cannot be detected by a UV detector without derivatization and therefore it is necessary to use reagents for derivatization. De Zwart et al. [2] derivatized a wide range of foods and analyzed betaine with liquid chromatography and UV detector using various columns. Slow et al. [3] have extracted betaine from various foods by using water and dichloromethane, and by derivatization of betaine with 2-naphthacyl trifluoromethanesulphonate. Hefni et al. [17] developed a simple HPLC-UV method in several different food matrices such as spinach, whole grain wheat flour, wheat, and sugar beet, with the help of derivatization on the cation exchange column. More recently, the same group has used the same method to analyze 14 cereal samples, representing different genera and cultivars [18]. Bruce et al. [19] and Ross et al. [11] performed the betaine analysis using LC-MS/MS (liquid chromatography with mass spectrometry) and HILIC (hydrophilic interactions liquid chromatography) column. Bruce et al. [19] developed the LC-MS/MS method for the analysis of 47 blood samples, 32 grains of cereals and cereal fractions, and 51 cereal products. Additionally, Ross et al. [11] analyzed betaine with LC-MS/MS in a wide range of commercially available cereals and cereal fractions. Also, recently Servillo et al. [4] have used LC-ESI-MS/MS for determination of different betaines present in commercial flours of cereals and pseudocereals. Instead of a conventional UV detector for the quantitative determination of betaine in order to avoid derivatization, the evaporative light scattering detector (ELSD) detector is used more recently and as a universal detector that provides a stable base line even in a gradient mode that can detect the majority non-volatile analytes. Shin et al. [20] have proposed a HILIC column in combination with an ELSD detector for betaine analysis. HILIC is an alternative to reverse phase chromatography, namely a type of normal phase chromatography, in which the stationary phase is polar but larger amounts of organic solvents can be used as a mobile phase as opposed to ordinary normal phase chromatography. Kojić et al. have used the HPLC-ELSD system using the HILIC column with isocratic mode of operation [21].

## 3. Cereal Grains as a Source of Betaine

Data on the distribution of betaine in various cereals and pseudocereals are scarce and there is definitely a lack of detailed study. Most data come from various studies that were focused on estimation of betaine dietary intake. Nevertheless, available studies report on wide variations in betaine content in cereals. Different types of cereals may have different amounts of betaine [22]. The following ranges were found by de Zwart et al. [2]: 270–1110 µg/g (dry solids) in wheat flour, and 200–1000 µg/g in oats. More detailed overview of betaine levels in various cereals and pseudocereals from different studies is displayed in Table 1. The displayed data showed that betaine content spanned in wide ranges within the studied grains. According to Corol et al. [22], betaine content in cereals varies depending on multiple factors including genotype and environmental differences such as geographical and/or year-to-year variations and their interactions with genotype. This study revealed a three-fold difference in glycine betaine content within bread wheat genotypes and a 3.8-fold difference across six environments. The highest glycine betaine levels were found in Hungarian wheat grains whereas the lowest in those grown in the UK [22]. Slow et al. [3] and de Zwart et al. [2] indicated that the level of betaine depends on the level of stress under which the crop grows. This is due to osmoprotectant and cryoprotectant function of betaine. For example, growth under drought can cause higher levels of betaine compared to well-watered crops.

Table 1. Betaine content in different samples of cereals and pseudocereals.

| Cereals and Pseudocereals | Betaine (µg/g Dry Weight) | References |
|---|---|---|
| Wheat (*Triticum aestivum*) | | |
| raw grain | 1150–1320 | [18] |
| | 490–574 | [23] |
| bran | 5047–5383 | [23] |
| | 2717 | [21] |
| | 2300–7200 | [3] |
| aleurone | 4538–6242 | [11] |
| germ | 3414 | [11] |
| wholegrain flour | 792 | [11] |
| | 730 * | [24] |
| | 604 | [19] |
| | 540 | [23] |
| refined flour | 718 * | [25] |
| | 700 * | [24] |
| | 415–593 | [21,23] |
| | 398 | [11] |
| | 180 * | [4] |
| | 141.2 | [19] |
| flour (not specified by origin) | 270–1110 | [2] |
| Wheat Emmer (*T. dicoccum*) | | |
| raw grain | 830–940 | [18] |
| refined flour | 195 * | [4] |
| Wheat Einkorn (*T. monococcum*) | | |
| refined flour | 367.3 * | [4] |
| Durum wheat (*T. durum*) | | |
| semolina | 1227 | [23] |
| | 483 | [21] |
| | 683 | [11] |
| refined flour | 253–303 | [23] |
| | 310 | [21] |
| wholegrain flour | 713 | [11] |
| | 245 * | [4] |
| Spelt wheat (*T. aestivum* ssp. spelta) | | |
| raw grain | 973–2723 | [23] |
| | 565–714 | [21] |
| wholegrain flour | 1296–1442 | [23] |
| | 1370–1430 | [18] |
| refined flour | 978 | [11] |
| | 522–593 | [23] |
| | 410 | [21] |
| Kamut wheat, Khorasan (*T. turgidum* ssp. turanicum) | | |
| raw grains | 1100 | [24] |
| Triticale (xTriticosecale) | | |
| raw grain | 986–1030 | [23] |
| Rye | | |
| raw grain | 2213 | [23] |
| | 1530–1760 | [18] |
| | 444 | [21] |
| bran | 1651 | [19] |
| refined flour | 310 * | [4] |
| wholegrain flour | 1500 * | [24] |
| | 1182 | [23] |
| | 986 | [21] |
| Barley | | |
| raw grain | 460 | [18] |
| raw grain from naked var. | 980 | [18] |
| wholegrain flour | 776–1023 | [23] |
| | 779 | [21] |
| refined flour | 250 * | [4] |
| flour from naked var | 424 | [21] |
| | 574 | [23] |
| pearled grain | 274 | [21] |

**Table 1.** *Cont.*

| Cereals and Pseudocereals | Betaine (µg/g Dry Weight) | References |
|---|---|---|
| Oats | | |
| raw grain | 280 | [18] |
| | 388 | [21] |
| raw grain from naked var. | 440 | [18] |
| wholegrain flour | 310 * | [24] |
| flour | 404–688 | [23] |
| | 53 * | [4] |
| bran | 200 * | [24] |
| | 190 | [11] |
| Maize | | |
| raw grain | 107–304 | [23] |
| | 175 | [21] |
| wholegrain meal | 120 * | [24] |
| degermed meal | 4 * | [24] |
| semolina | 3–22 | [11] |
| refined corn grits | 37 | [11] |
| flour, enriched | 20 * | [24] |
| refined flour | 2.1 * | [4] |
| bran | 184 | [21] |
| | 104 | [23] |
| | 46 * | [24] |
| flakes | 103–120 | [23] |
| | 7–9 | [11] |
| | n.d. | [21] |
| starch | n.d. | [21] |
| popped | 19 | [11] |
| | n.d. | [21] |
| Rice | | |
| grain | 1–5 | [11] |
| | n.d. | [21] |
| refined flour | 8.4 * | [4] |
| expanded | n.d. | [21] |
| starch | n.d. | [21] |
| Amaranth (*Amaranthus cruentus*) | | |
| raw grain | 7420 | [23] |
| | 680 * | [24] |
| | 646 | [11] |
| expanded grain | 669 | [23] |
| | 607 | [21] |
| flour | 895–1225 | [23] |
| | 871 | [21] |
| Proso millet | | |
| sample type not specified | 95–112 | [11] |
| dehulled grain | 281 | [23] |
| refined flour | 1320 * | [4] |
| Buckwheat | | |
| wholegrain flour | 108 | [23] |
| | 7–20 | [11] |
| refined flour | n.d. | [21] |
| groats, roasted | 10 * | [4] |
| | 26 * | [24] |
| Sorghum | | |
| refined flour | 425 * | [4] |
| Quinoa | | |
| | 6300 * | [24] |
| grains | 3042–4428 | [11] |
| | 610.8 * | [4] |

n.d. not detected; * result expressed on wet weight.

Among glutinous cereals, the highest content of betaine was found in the bran fraction of wheat grain (2300–7200 µg/g) and in the germ (3414 µg/g) (Table 1). In spelt wheat, higher upper betaine levels were detected in comparison to common wheat. Wholegrain spelt flour was much higher in betaine than the wholegrain flour of common wheat (Table 1). Wholegrain flours were mainly higher in betaine when compared to refined flours. Ross et al. [11] estimated that wholegrain flours and products were two to four times higher in betaine in comparison to the refined counterparts. Similar betaine content was found in flour from durum wheat and conventional wheat. In contrast, Ross et al. [11] reported higher levels of betaine in durum semolina in comparison to common, non-refined wheat.

The most abundant source of betaine was amaranth, a pseudocereal. Raw amaranth grains contained 7420 µg/g betaine which was the highest value determined in a single sample [23]. According to Ross et al. [11] and USDA database [24], quinoa can also be listed as an outstanding source of betaine, having been reported to contain 3930 µg/g and 6300 µg/g betaine, respectively.

## 4. Betaine Content in Cereal-Based Products

The betaine content in cereal products depends on the processing method. Two to four times lower betaine content were found in refined grain products compared to equivalent whole grain products [11]. Betaine content is notably dependent on the loss of bran fraction during processing. The higher the abrasion of aleurone layer, the lower the betaine content in the product. Outstanding betaine levels were determined in wheat bran, up to 7200 µg/g (Table 1). Likes et al. [25] analyzed the betaine contents in different milling streams and reported the lowest betaine level in the cleanest milling fractions. In the study of de Zwart et al. [2], a wide range of different foods was analyzed for betaine content and flour was denoted as an item high in betaine (730 µg/g), however it was not specified the type of flour, except that it was available from retail markets. Betaine ranges in bread, pasta, breakfast cereals and snacks are given in Table 2. As it can be seen, the variation within each product category is high due to versatility of ingredients in product formulations. In each product category, the highest betaine content was reported for wholegrain products or products containing bran or germ. Among breads, rye, spelt, and wholemeal breads were abundant in betaine. Moderate to high betaine contents were reported for pasta products, but it must be noted that mainly uncooked samples were analyzed (Table 2). Breakfast cereals are a mixture of cereal and non-cereal ingredients and the betaine content will depend on the contribution of each ingredient. In the study of Filipčev et al. [23], two samples of commercially available breakfast cereals were analyzed, one of which contained no detectable levels of betaine whereas the other had 471 µg/g (on dry solids). A similar concluded was made by Ross et al. [11] for muesli and muesli bars which were found to contain only low-to-moderate betaine levels. These products were mainly based on oats and contained other low-betaine ingredients such as dried fruits. In contrast to Ross et al. [11], the USDA data [24] report on much wider span of betaine in breakfast cereals, from 7 µg/greaching to as much as 3600 µg/g (on wet weight) betaine.

**Table 2.** Betaine content in various grain-based products.

| Product | Betaine Content (µg/g Dry Weight) | References |
|---|---|---|
| Bread | | |
| rye bread | 855–1377 | [11] |
| wholegrain spelt | 913 | [11] |
| wholemeal | 670–790 | [3] |
| wholegrain | 499–781 | [11] |
| | 560–620 | [3] |
| multigrain | 247–678 | [11] |
| white (refined) | 360–520 | [3] |
| | 174–287 | [11] |
| various (white, sourdough) | 310–590 * | [24] |
| | 380 * | [24] |
| | 579 | [19] |
| wheat tortilla | 311 | [11] |

Table 2. *Cont.*

| Product | Betaine Content (μg/g Dry Weight) | References |
|---|---|---|
| Pasta | | |
| wholegrain wheat pasta | 710–1286 | [11] |
| | 375 | [19] |
| pasta, not specified | 480–1350 | [2] |
| refined wheat pasta | 628–706 | [11] |
| refined wheat (*T. aestivum*) pasta, uncooked | 253 | [21] |
| durum wheat pasta, uncooked | 188 | [21] |
| one–egg spelt pasta | 243–516 | [11] |
| barley pasta | 211 | [11] |
| noodles with egg, enriched, uncooked | 1300 * | [24] |
| noodles with egg, enriched, cooked | 190 * | [24] |
| refined couscous | 691 | [11] |
| bulghur | 1311 | [11] |
| cooked bulghur | 830 * | [24] |
| Breakfast cereals | | |
| ready-to-eat wheat germ, toasted, plain | 4100 * | [24] |
| ready-to-eat wheat bran, toasted | 3200 * | [24] |
| wholegrain rye flakes | 1640 | [11] |
| wholegrain wheat-based cereals | 732–915 | [11] |
| wholegrain oat and wheat-based muesli | 310 | [11] |
| wholegrain oat-based muesli | 117–226 | [11] |
| breakfast cereals, not specified | 180–300 | [21] |
| muesli bar | 171 | [11] |
| wholegrain porridge oats | 128–167 | [11] |
| extruded whole grain oat cereals | 73–91 | [11] |
| cereal bar | 74–75 | [11] |
| various ready-to-eat cereals | 7–3600 * | [24] |
| Snacks, cookies, crackers, crispbread, cakes, pastry | | |
| wholegrain rye crispbread | 1428–1527 | [11] |
| frozen, read-to-eat pancakes | 690–720 * | [24] |
| wholegrain wheat crackers | 293–649 | [11] |
| crackers, classic, saltines, cheese | 340–580 * | [24] |
| wholegrain wheat rusks | 556–564 | [11] |
| wholegrain wheat muffin | 437–501 | [11] |
| various commercial cakes | 190–480 * | [24] |
| wholegrain wheat biscuit | 425 | [21] |
| Graham cookies | 390 * | [24] |
| doughnuts | 270–380 * | [24] |
| English muffins | 220–360 * | [24] |
| extruded spelt | 308 | [21] |
| refined wheat crackers | 258–332 | [11] |
| digestive biscuit | 271–309 | [11] |
| apple pie, commercial | 160 * | [24] |
| biscuit | 4–144 | [11] |
| Danish pastry, fruit enriched | 140 * | [24] |
| plain Danish pastry | 81 * | [24] |

* Result expressed on wet weight.

## 5. Betaine Content in Gluten-Free Cereal Products

Gluten-free products have been generally recognized to be low in betaine content [11,19]. In the majority of commercially available gluten-free products, a very low level of betaine (<50 μg/g) was observed [11]. Table 3 lists the betaine levels reported for commercial gluten-free products from several studies. In the bread and biscuits category, betaine levels ranged from non-detectable to 107 μg/g. Similar findings were reported by Kojić et al. [21], who also found that gluten-free samples (starch, corn extrudates, pasta, cornflakes, and rice) contained no detectable levels of betaine. Gluten-free cereals contained much lower amounts of betaine in comparison to glutenous cereals: corn had

107–304 µg/g betaine [23]; teff and millet between 50–150 µg/g [11], proso millet 280 µg/g [23]. Buckwheat is a frequent ingredient in gluten-free products. According to Ross et al. [11], buckwheat was among those ingredients low in betaine (<20 µg/g) although as high as 390 µg/g betaine was found in buckwheat uncooked pasta (Table 3).

As mentioned earlier, gluten-free ingredients with appreciable amounts of betaine are amaranth and quinoa. Amaranth grain was reported to contain 646–680 µg/g betaine and a remarkable figure of 7420 µg/g betaine in a single sample of raw grains (Table 1). Processed amaranth contained 817–1225 µg/g of betaine (flour) and 669 µg/g (expandate) (Table 1).

In order to increase betaine levels in gluten-free products and consequently improve dietary intake of betaine among people adhereing to gluten-free and vegan diets, the incorporation of amaranth, quinoa, proso millet, and buckwheat as base ingredients into gluten-free products as well as their enrichment with sugar beet molasses was proposed [11]. Sugar beet molasses can remarkably increase betaine content in some baked products, even when used at fortification levels that do not compromise the sensory properties [26]. It was reported a 43× increase in betaine content of molasses-enriched plain biscuits in comparison to the control biscuit (without molasses) [23]. In gluten-free cookies enriched with molasses at 30% (flour basis), the betaine level was raised ≈64 times [26].

When considering fortification of products with betaine, the challenge is to achieve sufficient delivery of betaine for the claim for lowering blood homocysteine (500 mg/portion). A trial to incorporate betaine in the range from 0.5% to 3% (flour basis) into the formulation of gluten-free biscuits revealed that they were capable of providing from 280 to 1370 mg of betaine per 100 g [27]. The highest fortification level (3%) significantly increased the biscuit spread and contributed to perceiving a weak aftertaste described as burning-like sensations of the tongue and palate which might be due to weak acid reaction of betaine [27]. Rising betaine doses improved color vividness in the biscuits [27]. Similar results were observed in the case of fortifying plain wheat cookies with betaine [28].

Table 3. Betaine content in gluten-free products.

| Product | Betaine Content (µg/g Dry Weight) | References |
|---|---|---|
| Bread and biscuits | | |
| gluten-free crispbread | 9–107 | [11] |
| savory biscuits | n.d.–104 | [23] |
| wholegrain gluten-free bread | 12–68 | [11] |
| oatmeal biscuits | 3 | [11] |
| gluten-free flour enriched with fibers | 1 | [11] |
| sweet biscuits | n.d. | [21] |
| flour mixture for gluten-free bread | n.d. | [21] |
| gluten-free cookies with almonds, crackers, salty sticks | n.d. | [21] |
| expanded maize | n.d. | [21] |
| Pasta | | |
| buckwheat pasta, uncooked | 390 | [23] |
| | 382 | [11] |
| | 175 | [21] |
| maize-based pasta | 2–20 | [11] |
| maize and rice-based pasta, uncooked | n.d. | [21] |
| rice-based pasta, uncooked | n.d. | [21] |
| Breakfast cereals and related products | | |
| soy bran | 182 | [21] |
| unseasoned popcorn | 19 | [11] |
| cornflakes | 14 | [11] |
| buckwheat flakes | 10 | [11] |
| rice-based breakfast cereals | 4–5 | [11] |
| expanded rice | n.d. | [21] |

n.d. not detected.

## 6. Stability of Betaine in Grain-Based Products

Betaine is known to be a thermostable compound which survives the severe treatment during sugar beet processing (extracting with water, treatment with $CaOH_2$ and $CO_2$, concentration, crystallization) and almost quantitatively accumulates in molasses [29]. Pure anhydrous betaine decomposes at > 245 °C. Since food processing practices do not employ such high temperatures, betaine losses caused by food thermal treatments were initially not expected [30]. However, some data suggest that certain cooking and baking losses of betaine may exist in spite of its thermostability in the pure form. Being a water soluble compound with a small molecule, it is not unlikely that some betaine losses will occur, depending on the type of food processing and cooking.

Only few studies exist that deal with the stability of betaine in food during processing. De Zwart et al. [2] compared the average betaine content in various food, before and after cooking. They concluded that the level of betaine varied widely, depending on the food and cooking method. The lowest losses (10–14%) were observed with microwave cooking of vegetables (frozen peas, silverbeet) and the highest losses with boiling (43–73%) [2]. In the case of cereal-based food, high betaine losses (76–84%) were detected during pasta boiling which could be attributed to dissolution of betaine in cooking water and its removal upon water draining [2]. Betaine reduction of ≈85% between uncooked and cooked noodles was reported in the USDA database [24] (Table 2). Similar results were confirmed by Ross et al. [11] in cooked pasta and noodles. During baking of scones, de Zwart et al. [2] determined a 17% betaine loss. Similar betaine losses were found after baking gluten-free biscuits fortified with betaine at a 0.5–3.0% level [27]. Somewhat higher baking losses in betaine were reported by Filipčev et al. [28] ranging from 17% to 28.6% in wheat biscuits fortified with betaine at 0.5–3.0% level. Very high betaine losses (>90%) were observed after baking betaine-enriched bread [31]. It was assumed that this loss could be partly due to betaine consumption by baker's yeast throughout dough fermentation since yeast can use betaine as a source of nitrogen.

During the preparation of extruded snack products enriched with betaine, the influence of extrusion cooking parameters like screw speed, feed flow rate, and feed moisture on the betaine content was analyzed [32]. The most significant influences on betaine content were feed rate and feed moisture content. Under the most extreme conditions applied during the extrusion process betaine losses were from 50–60% [32].

In some cases, an increase in betaine could be observed after thermal treatment as reported for fried and baked falafel [2] and for oatmeal cooked in a microwave oven [11]. The increases were 9, 14, and 31%, respectively. Ross et al. [11] suggested that a plausible explanation of this phenomenon could be the liberation of betaine from food matrix or betaine synthesis throughout heating. So far, it has not been reported that betaine is capable of forming any bonds with matrix components.

## 7. Conclusions

Comparison of betaine levels in cereals from different studies showed that cereals are good sources of betaine. Wheat bran and germs were the most abundant wheat fractions. There were large differences in the betaine contents of different cereals. The wholegrain flours from spelt, rye, and barley showed higher betaine levels in comparison to that of common wheat. Non-glutinous cereals are generally moderately to very low in betaine. The best gluten-free sources of betaine are amaranth and quinoa.

Cereal grain processing may lead to lowering of betaine content, especially if removal of aleurone layers is included. Thermal treatment of cereal products also provokes certain loss of betaine, in spite of its thermal stability on food processing temperatures. Losses are very high if processing involves water removal after cooking or boiling, since betaine is soluble in water. Very high losses were observed during baking of betaine-enriched bread, implying that fermentation by baker's yeast may be one of the causes but future research is needed to understand the possible mechanisms.

*Foods* **2018**, *7*, 49

Fortification of grain-based food with sugar beet molasses even at low or moderate levels considerably raises their betaine content and may be considered as a possible way to increase the functionality of the products.

**Acknowledgments:** The authors declare the financial support from the Ministry of Education, Science and Technological Development, Serbia, grant III 46005. No funds were received to cover publication costs.

**Author Contributions:** Nebojša Ilić and Marija Bodroža-Solarov conceived and reviewed the manuscript; Jovana Kojić contributed to data collection, analysis, interpretation, and to writing the paragraph dealing with methods for betaine analysis; Jelena Krulj contributed to data collection, analysis, and interpretation; Bojana Filipčev performed the background research and wrote the manuscript.

**Conflicts of Interest:** The authors declare no conflict of interest.

## References

1. Yancey, P.H.; Clark, M.E.; Hand, S.C.; Bowlus, R.D.; Somero, G.N. Living with stress: Evolution of osmolyte systems. *Science* **1982**, *217*, 1214–1222. [CrossRef] [PubMed]
2. De Zwart, F.J.; Slow, S.; Payne, R.J.; Lever, M.; George, P.M.; Gerrard, J.A.; Chambers, S.T. Glycine betaine and glycine betaine analogues in common foods. *Food Chem.* **2003**, *83*, 197–204. [CrossRef]
3. Slow, S.; Donaggio, M.; Cressey, P.J.; Lever, M.; George, P.M.; Chambers, S.T. The betaine content of New Zealand foods and estimated intake in the New Zealand diet. *J. Food Compos. Anal.* **2005**, *18*, 473–485. [CrossRef]
4. Servillo, L.; D'Onofrio, N.; Giovane, A.; Casale, R.; Cautela, D.; Ferrari, G.; Castaldo, D.; Balestrieri, M.L. The betaine profile of cereal flours unveils new and uncommon betaines. *Food Chem.* **2018**, *239*, 234–241. [CrossRef] [PubMed]
5. Craig, S.A. Betaine in human nutrition. *Am. J. Clin. Nutr.* **2004**, *80*, 539–548. [CrossRef] [PubMed]
6. Olthof, M.R.; Van Vliet, T.; Boelsma, E.; Verhoef, P. Low dose betaine supplementation leads to immediate and long term lowering of plasma homocysteine in health men and women. *J. Nutr.* **2003**, *133*, 4135–4138. [CrossRef] [PubMed]
7. Steenge, G.R.; Verhoef, P.; Katan, M.B. Betaine supplementation lowers plasma homocysteine in healthy men and women. *J. Nutr.* **2003**, *133*, 1291–1295. [CrossRef] [PubMed]
8. Hoffman, J.R.; Ratamess, N.A.; Kang, J.; Rashti, S.L.; Faigenbaum, A.D. Effect of betaine supplementation on power performance and fatigue. *J. Int. Soc. Sports Nutr.* **2009**, *6*, 7–17. [CrossRef] [PubMed]
9. Melnyk, S.; Fuchs, G.J.; Schulz, E.; Lopez, M.; Kahler, S.G.; Fussell, J.J.; Bellando, J.; Pavliv, O.; Rose, S.; Seidel, L.; et al. Metabolic imbalance associated with methylation dysregulation and oxidative damage in children with autism. *J. Autism Dev. Disord.* **2012**, *42*, 367–377. [CrossRef] [PubMed]
10. Jill James, S.; Melnyk, S.; Jernigan, S.; Cleves, M.A.; Halsted, C.H.; Wong, D.H.; Cutler, P.; Bock, K.; Boris, M.; Bradstreet, J.J.; et al. Metabolic endophenotype and related genotypes are associated with oxidative stress in children with autism. *Am. J. Med. Genet.* **2006**, *141B*, 947–956. [CrossRef] [PubMed]
11. Ross, A.B.; Zangger, A.; Guiraud, S.P. Cereal foods are the major source of betaine in the Western diet—Analysis of betaine and free choline in cereal foods and updated assessments of betaine intake. *Food Chem.* **2014**, *145*, 859–865. [CrossRef] [PubMed]
12. Chu, D.M.; Wahlgvist, M.L.; Chang, H.Y.; Yeh, N.H.; Lee, M.S. Choline and betaine food sources and intakes in Taiwanese. *Asia Pac. J. Clin. Nutr.* **2012**, *21*, 547–557. [PubMed]
13. Konstantinova, S.V.; Tell, G.S.; Vollset, S.E.; Ulvik, A.; Drevon, C.A.; Ueland, P.M. Dietary patterns, food groups, and nutrients as predictors of plasma choline and betaine in middle-aged and elderly men and women. *Am. J. Clin. Nutr.* **2008**, *88*, 1663–1669. [CrossRef] [PubMed]
14. European Commission (EC). Commission Regulation (EU) No. 432/2012 of 16 May2012 establishing a list of permitted health claims made on foods, other than those referring to the reduction of disease risk and to children's development and health. *Off. J. Eur. Union* **2012**, *136*, 1–40. Available online: https://www.fsai.ie/uploadedFiles/Reg432_2012.pdf (accessed on 15 August 2015).
15. Atkinson, W.; Slow, S.; Elmslie, J.; Lever, M.; Chambers, S.; George, P. Dietary and supplementary betaine: Effects on betaine and homocysteine concentrations in males. *NMCD* **2009**, *19*, 767–773. [CrossRef] [PubMed]

16. Saarinen, M.T.; Kettunen, H.; Pulliainen, K.; Peuranen, S.; Tiihonen, K.; Remus, J. A novel method to analyze betaine in chicken liver: Effect of dietarymbetaine and choline supplementation on the hepatic betaine concentration in broiler chicks. *J. Agric. Food Chem.* **2001**, *49*, 559–563. [CrossRef] [PubMed]

17. Hefni, M.; McEntyre, C.; Lever, M.; Slow, S. Validation of HPLC-UV methods for the quantification of betaine in foods by comparison with LC-MS. *Food Anal. Methods* **2016**, *9*, 292–299. [CrossRef]

18. Hefni, E.M.; Schaller, F.; Witthöft, M.C. Betaine, choline and folate content in different cereal genotypes. *J. Cereal Sci.* **2018**, *80*, 72–79. [CrossRef]

19. Bruce, S.J.; Guy, P.A.; Rezzi, S.; Ross, A.B. Quantitative measurement of betaine and free choline in plasma, cereals and cereal products by isotope dilution LC-MS/MS. *J. Agric. Food Chem.* **2010**, *58*, 2055–2061. [CrossRef] [PubMed]

20. Shin, H.D.; Suh, J.H.; Kim, J.H.; Lee, H.Y.; Eom, H.Y.; Kim, U.Y.; Youm, J.R. Determination of betaine in Fructus Lycii using hydrophilic interaction liquid chromatography with evaporative light scattering detection. *Bull. Korean Chem. Soc.* **2012**, *33*, 553–558. [CrossRef]

21. Kojić, J.; Krulj, J.; Ilić, N.; Lončar, E.; Pezo, L.; Mandić, A.; Bodroža-Solarov, M. Analysis of betaine levels in cereals, pseudocereals and their products. *J. Funct. Foods* **2017**, *37*, 157–163. [CrossRef]

22. Corol, D.I.; Ravel, C.; Raksegi, M.; Bedo, Z.; Charmet, G.; Beale, M.H.; Ward, J.L. Effects of genotype and environment on the contents of betaine, choline, and trigonelline in cereal grains. *J. Agric. Food Chem.* **2012**, *60*, 5471–5481. [CrossRef] [PubMed]

23. Filipčev, B.V.; Brkljača, J.S.; Krulj, J.A.; Bodroža-Solarov, M.I. The betaine content in common cereal-based and gluten-free food from local origin. *Food Feed Res.* **2015**, *42*, 129–137. [CrossRef]

24. Patterson, K.Y.; Bhagwat, S.A.; Williams, J.R.; Howe, J.C.; Holden, J.M. USDA Database for the Choline Content of Common Foods—Release 2. Available online: http://www.ars.usda.gov/SP2UserFiles/Place/12354500/Data/Choline/Choln02.pdf (accessed on 5 January 2018).

25. Likes, R.; Madl, R.L.; Zeisel, S.H.; Craig, S.A.S. The betaine and choline content of a whole wheat flour compared to other mill streams. *J. Cereal Sci.* **2007**, *46*, 93–95. [CrossRef] [PubMed]

26. Filipčev, B.; Mišan, A.; Šarić, B.; Šimurina, O. Sugar beet molasses as an ingredient to enhance the nutritional and functional properties of gluten-free cookies. *Int. J. Food Sci. Nutr.* **2016**, *67*, 249–256. [CrossRef] [PubMed]

27. Filipčev, B.; Krulj, J.; Brkljača, J.; Šimurina, O.; Jambrec, D.; Bodroža-Solarov, M. Fortification of gluten-free biscuits with betaine. In Proceedings of the 8th International Congress FLOUR-BREAD'15 and 10th Croatian Congress of Cereal Technologists, Opatija, Croatia, 29–30 October 2015; Koceva-Komlenić, D., Ed.; Faculty of Food Technology Osijek, University of Osijek: Osijek, Croatia, 2016; pp. 92–98. Available online: http://www.ptfos.unios.hr/joomla/znanost/flour-bread/images/PDF/proceedings_FB15_web.pdf (accessed on 16 October 2016).

28. Filipčev, B.; Krulj, J.; Kojić, J.; Šimurina, O.; Bodroža Solarov, M.; Pestorić, M. Quality attributes of cookies enriched with betaine. In Proceedings of the III International Congress "Food Technology, Quality and Safety-FOODTECH", Novi Sad, Serbia, 25–27 October 2016; Đuragić, O., Ed.; Institute of Food Technology, University of Novi Sad: Novi Sad, Serbia; pp. 46–51.

29. Šušić, S.; Sinobad, V. Ispitivanja u cilju unapređenja industrije šećeraJugoslavije. *Hem. Ind.* **1989**, *43* (Suppl. 1–2), 10–21.

30. The Scientific Panel on Dietetic Products, Nutrition and Alergies. Opinion on the scientific panel on dietetic products, nutrition and allergies on a request from the Commission related to an application concerning the use of betaine as a novel food in the EU. *EFSA J.* **2005**, *191*, 1–17.

31. Filipčev, B.; Šimurina, O.; Brkljača, J.; Krulj, J.; Bodroža-Solarov, M.; Popov, S. Nutritional quality and baking performance of bread enriched with betaine. In Proceedings of the 11th Symposium "Novel Technologies and Economic Development", Leskovac, Serbia, 23–24 October 2015; Lazić, M., Ed.; Faculty of Technology in Leskovac, University of Niš: Leskovac, Serbia; pp. 83–88.

32. Kojić, J.S.; Ilić, N.M.; Kojić, P.S.; Pezo, L.L.; Banjac, V.V.; Krulj, J.A.; Bodroža-Solarov, M.I. Twin-Screw Extrusion of Spelt Flour Enriched with Betaine—Multi-Objective Optimization Approach. In preparation.

*foods*

MDPI

Review

# How Safe Is Ginger Rhizome for Decreasing Nausea and Vomiting in Women during Early Pregnancy?

Julien Stanisiere, Pierre-Yves Mousset and Sophie Lafay *

GYNOV SAS, 5 rue Salneuve, 75017 Paris, France; j.stanisiere@gynov.com (J.S.); py.mousset@jenwin.fr (P.-Y.M.)
* Correspondence: s.lafay@gynov.com; Tel.: +33-983-395-434

Received: 2 February 2018; Accepted: 19 March 2018; Published: 1 April 2018

**Abstract:** Ginger, *Zingiber officinale* Roscoe, is increasingly consumed as a food or in food supplements. It is also recognized as a popular nonpharmacological treatment for nausea and vomiting of pregnancy (NVP). However, its consumption is not recommended by all countries for pregnant women. Study results are heterogeneous and conclusions are not persuasive enough to permit heath care professionals to recommend ginger safely. Some drugs are also contraindicated, leaving pregnant women with NVP with few solutions. We conducted a review to assess effectiveness and safety of ginger consumption during early pregnancy. Systematic literature searches were conducted on Medline (via Pubmed) until the end of December 2017. For the evaluation of efficacy, only double-blind, randomized, controlled trials were included. For the evaluation of the safety, controlled, uncontrolled, and pre-clinical studies were included in the review. Concerning toxicity, none can be extrapolated to humans from in vitro results. In vivo studies do not identify any major toxicities. Concerning efficacy and safety, a total of 15 studies and 3 prospective clinical studies have been studied. For 1 g of fresh ginger root per day for four days, results show a significant decrease in nausea and vomiting and no risk for the mother or her future baby. The available evidence suggests that ginger is a safe and effective treatment for NVP. However, beyond the ginger quantity needed to be effective, ginger quality is important from the perspective of safety.

**Keywords:** pregnancy; *Zingiber officinale* R; ginger; NVP; toxicity; safety; adverse effects; food supplement; CAM

## 1. Introduction

According to the World Health Organization, the growth and expansion of traditional and complementary medicine (T&CM) products is worldwide phenomenon. These days, this sector plays a significant role in the economic development of number of countries [1]. The growing mistrust of the side effects of pharmaceuticals products coupled with the desire for more traditional medicines perceived as natural and safe by consumers, may partially explain the increase in the use of herbal remedies.

In Europe, many herbal remedies are sold as food supplements and meet Directive 2002/46/EC [2]. Regarding botanicals, particularly, the European Food Safety Authority (EFSA) published, in 2012, a compendium of botanicals and associated substances of concern. The objective is to have a guideline for the evaluation of specific ingredients in food supplements, identifying the compound(s) to monitor [3]. In parallel, positive and/or negative lists of botanicals are published by authorities in different European countries [4–7] to control and guarantee the quality and traceability of used botanical ingredients in food supplements. These regulatory approaches aim to protect consumer health by ensuring that Complementary and Alternative Medicine (CAM) are safe and of high quality.

More than 100 million Europeans use regularly CAM, and the prevalence varied from 5.9 to 48.3% [8]. However, women in the middle-age, tertiary educated are typical consumers. In France [9], women consume more food supplements than men. This is also true in Belgium [10], Australia, and the USA [11]. Despite higher consumption by women in Europe, there are variations by country [12]. The differences in consumption by country and gender are summarized in Table 1.

**Table 1.** Comparison of food supplement consumption between men and women by country [12].

| Country | Men (%) | Women (%) |
|---|---|---|
| France | 17–21.6 | 26.3–28.5 |
| Belgium | 11–14 | 22.5–30 |
| Australia | 34.9 | 50.3 |
| USA | 45 | 58 |
| Greece | 2 | 6.7 |
| Spain | 5.9 | 12.1 |
| Italy | 6.8 | 12.6 |
| Germany | 20.7 | 27 |
| The Netherlands | 16 | 32.1 |
| UK | 36.3 | 47.5 |
| Denmark | 51 | 65.8 |
| Sweden | 30.5 | 42.4 |

Not only women, but also pregnant women, consume food supplements. In the United States, 36.7% of pregnant women from the ages of 19 to 49 reported using CAM in the last year compared to 40.7% of non-pregnant women [13]. In the UK and Australia, 57.8% and 52% of pregnant women, respectively, had used an herbal remedy during pregnancy [14,15].

Some studies showed that the women had a positive opinion on the safety of herbal remedies during pregnancy [16,17]. Moreover, a minimum of one-third of the healthcare professionals are willing to recommend the use of CAM to pregnant women [18,19], and the majority (69.2%) agreed that there was some value in CAM use during pregnancy [20]. Nevertheless, the safety of CAM was also a key concern. Despite this, documentation on the safety and efficacy of many herbs used during pregnancy is limited. Whereas most Member States update and regularly implement regulations relative to herbal substances/products according to the most recent scientific evaluations, few toxicological data coming from studies on pregnant women are available.

A multinational, cross-sectional study [18] identified 126 specific herbal medicines used by 2379 women. They were classified on three categories: safe, caution, and contraindicated, and their consumption was observed. Women used herbal medicine mainly classified as safe for pregnancy use. However, a difference between regions was observed (Figure 1). This study shows that there is still work to be done in informing pregnant women. Information confirmed by Pallivalappila et al. [19], 61% (*n* = 127) of dietary supplement users during pregnancy and 44% (*n* = 54) of non-users responded that CAM should be available through the NHS (the publicly-funded healthcare system in Scotland).

One of the most popular botanical remedies during pregnancy is ginger (*Zingiber officinale* Roscoe) [14,18,21]. Ginger is an Asian native plant. Its aromatic rhizome is used as a spice, but also in traditional medicine since ancestral times. Ginger belongs to the official pharmacopoeias of different countries, including Austria, China, Egypt, India, United Kingdom, Japan, Switzerland, and the Netherlands.

For around 10 years, ginger imports in Europe have increased significantly [22], and many foods and food supplements have appeared on the market. Most of these food supplements are dedicated to pregnant women (Table 2).

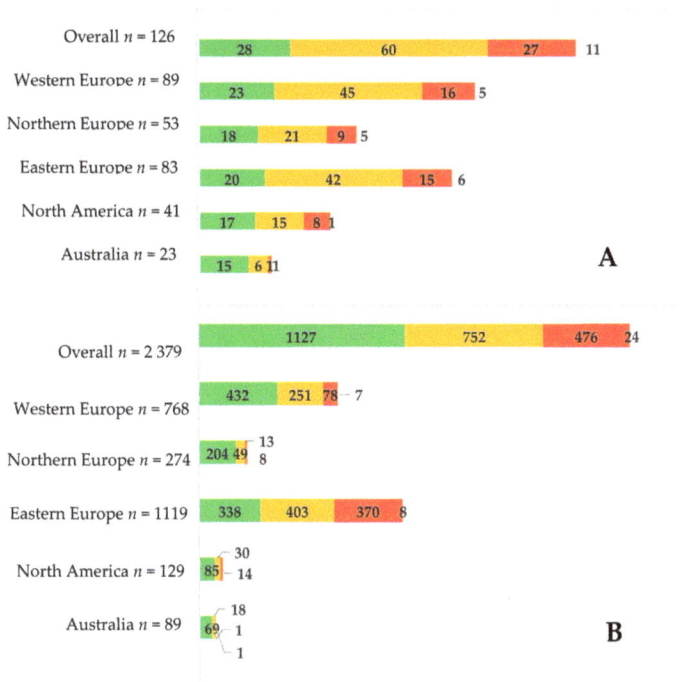

**Figure 1.** Herbal medicine used (**A**) and number of women who used them (**B**) by safety classification (green = safe; orange = caution; red = contraindicated; white = unknown). From Kennedy et al. [18].

**Table 2.** Sample of food supplements with ginger sold around the world; composition and claims.

| Country | Composition | Daily Dose | Claims |
|---|---|---|---|
| Belgium | Cellulose; standardized ginger extract 50 mg with 10% gingerols (equal to 500 mg of powder); calcium phosphate; silicon dioxide, magnesium stearate; hypromellose, titanium dioxide; talc; glycerol | 2/day for pregnant women | Helps you regain optimal digestive balance in the following situations: - in case of overeating - in case of unusual diets - when agitated or on trips - too much stress Can be used in pregnant women who wish to return to digestive well-being during pregnancy |
| | Ginger bio extract (6 × concentrated) 200 mg | 1/day | Digestive comfort |
| | 500 mg of ginger rhizome powder and 5 mg of ginger rhizome extract per caps | 1/day | Sexual fatigue; nausea in pregnant women; travel sickness |
| | Ginger rhizome extract: 67 mg, eq. to 1 g of ginger rhizome, magnesium carbonate, vitamin B6 | 1/day | Ginger contributes to the normal functioning of the stomach in case of early pregnancy (Claim under consideration (EFSA)) Magnesium contributes to a reduction of tiredness and fatigue Vitamin B6 contributes to the regulation of hormonal activity |

**Table 2.** *Cont.*

| Country | Composition | Daily Dose | Claims |
|---------|-------------|------------|--------|
| France | Organic ginger rhizome powder: 250 mg | 4/day | Nausea in pregnant women |
| | Ginger rhizome powder (*Zingiber officinale* R): 365 mg/hard caps | 4/day | Lower bowel contractions and digestive acids; prevent motion-induced nausea and vomiting |
| | Organic ginger rhizome extract (*Zingiber officinale* R): 200 mg | 2/day | Ginger contributes to the normal functioning of the stomach in case of early pregnancy |
| | Organic ginger rhizome extract (*Zingiber officinale* R): 200 mg | 5/day | Travel sickness |
| | *Zingiber officinalis* R (roots): 230 mg. | 6/day | Helps to support the digestion/contributes to the normal function of intestinal tract<br>Travel sickness<br>Joint mobility<br>Libido in men |
| | Sorbitol, rhizome ginger extract (*Zingiber officinale* R), natural orange flavor, sodium starch glycolate, microcrystalline cellulose, natural lemon flavor, magnesium stearate, silica, sucralose. | 2/day | - |
| | Ginger, lemon, B6, magnesium, iron, B9 | 5 biscuits/day | - |
| | B6, folates, cocoa, strawberry pulp, ginger | 1/day | - |
| | Ginger (*Zingiber officinale* Roscoe), cinnamon (*Cinnamomum verum* J. Presl), fennel (*Foeniculum vulgare* Mill.), lemon (*Citrus limon* (L.) Burm f.), Tumeric (*Curcuma longa* L.), hibiscus (*Hibiscus sabdariffa* L.), liquorice (*Glycyrrhiza glabra* L), natural lemon flavor and natural vanilla flavor | 1–3 sachets /day | - |
| | Cellulose, silicon dioxide, magnesium stearate, standardized ginger extract (*Zingiber officinale* R) 50 mg; HPMC, titanium dioxide, talc, copper complexes of chlorophyllins | Pregnant women: 2/day<br>Children from 6 to 11 years old: 1 to 4/day<br>Adults and Children >12 years old: 2 to 8/day | Nausea in pregnant women; travel sickness<br>Ginger contributes to the normal functioning of the stomach (Claim under consideration (EFSA)) |
| | Cellulose, standardized ginger extract (10% of gingerols) 50 mg eq. to 500 mg of ginger powder; calcium phosphate; silicon dioxide; magnesium stearate; HMPC; titanium dioxide, talc, glycerol. | 2/day for pregnant women | Helps to support the digestion/contributes to the normal function of intestinal tract/contributes to the normal functioning of the stomach in case of early pregnancy or travel sickness<br>For adults and children aged 12 years and above |
| | Microcrystalline cellulose, standardized ginger extract 50 mg, eq. to 500 mg of rhizome powder, silicon dioxide, fatty acid magnesium salts, hypromellose, titanium dioxide, talc, glycerol | 2 hard caps 3 times per day | Helps to support the digestion<br>Contributes to decrease discomforts in case of travel nausea, pregnant women nausea, or chemotherapy |
| | Bach flowers, essential oils, and plant extracts | 5 sprays if nauseous | Quick-acting oral spray against discomfort, apprehension and transport. |
| | Ginger root extract (160 mg of extract, 16 mg of gingerols); lemon balm extract (100 mg) | 4/day | Stomach; antiemetic; pregnant women |
| Italy | Standardized root extract: 300 mg (5% gingerols) Whole root pulverized: 1500 mg | 2/day | Regulate the gastro-intestinal motility and gas elimination of in case of nausea<br>Promotes joint function and counteracts the localized states of tension<br>Counter menstrual cycle disorders |
| | Hydro-alcoholic standardized ginger extract (5% of gingerols) | 2/day | Reducing nausea, gas, bloating and intestinal spasms |
| Poland | Ginger rhizome extract 75 mg (eq. to 300 mg of powder); Vitamin B6: 0.5 mg | 3/day | Adults and children poorly tolerating travel by vehicles and craft: cars, buses, trains, ships, planes |
| UK | Ginger extract 55 mg eq. to 1.1 g of root powder | 1/day | Helps calm a queasy stomach; an excellent stomach soother; excellent travel companion |
| | Ginger extract 120 mg eq. to 14 g of root powder (24 mg of gingerols) (120:1 extract) | 1/day | Do not take if pregnant or breastfeeding |
| | Ginger extract 138 mg eq. to 550 mg | 2/day | - |
| | Ginger root 500 mg | 2/day | - |
| | Ginger root extract (*Zingiber officinale* R) 500 mg, vitamin B6 25 mg, vitamin D 200 IU, calcium (carbonate and lactate) 240 mg, red raspberry leaf 25 mg, mint leaf 25 mg<br>*Citrus aurantifolia* 10 mg<br>Contains no artificial colors or flavors. | 2/day | |
| US | Ginger root powder | 2/day | - |
| | Ginger root powder | 2/day | May soothe an upset stomach and support digestion<br>Promotes a healthy inflammation response<br>May help reduce nausea, vomiting and dizziness |
| | Ginger root powder | 2/day | Helps calm a queasy tummy an excellent stomach soother Excellent travel companion |

Whereas ginger rhizome consumption is described for Nausea and Vomiting of Pregnancy (NVP) in different monographs [23–25], it is not recommended in others as a precautionary measure [26]. In the same way, ginger rhizome consumption by pregnant women is tolerated or authorized in several countries (France, Belgium) or forbidden in others (Finland, Russia). Without scientific evidence established in the concerned population, clear recommendations, and harmonized regulations, it is difficult for healthcare professionals to provide safe advice.

This review focuses on the efficacy for NVP of ginger and the safety of its use in pregnant women. The main goal is to have an objective treatment of safety, so non-clinical and clinical data were analyzed to present and discuss factual information to healthcare professionals.

## 2. Generality

### 2.1. Zingiber Officinale Roscoe

Ginger, *Zingiber officinale* Roscoe is originated from Asia. It is a plant of the genus *Zingiber* and of the family Zingiberaceae, whose rhizome is used worldwide in cooking and traditional medicine. It is considered as premium spices, like cardamom and turmeric. Ginger was first cultivated in the Asian subcontinent, probably in South East Asia [27]. It migrated to Europe by Greek and Roman times and was used as digestive aids wrapped in bread. Ginger was then incorporated into bread and confections. In the 1600s, the Spanish established ginger plantations in Jamaica and in the 19th century some physicians used it to induce sweating, to improve the appetite or decrease the nausea [28].

Ginger is cultivated in the tropical regions from both hemispheres and India is the largest producer (32.75% of the world's production), followed by China (21.41%) and Nigeria (12.54%) [23,29].

### 2.2. Nutritional Composition and Chemical Composition

The main ginger constituents are starch (up to 50%), lipids (6 to 8%), proteins, and inorganic compounds [25,30]. When referring to the USDA National Nutrient Database for Standard Reference and DTU Fødevaredatabanken, raw ginger root is a source of potassium (415 mg/100 g). When referring to the ANSES Table Ciqual 2017 [31], data are very different: ginger powder is a source of phosphorus (168 mg/100 g) and is rich in magnesium (214 mg/100 g), potassium (1320 mg/100 g), manganese (33.3 mg/100 g), zinc (3.64 mg/100 g), iron (19.8 mg/100 g), and niacin (9.62 mg/100 g) (Table 3). These differences between nutrient databases showed high quantitative variations that could be explained by the plant variability itself but also by used analytical methods which can differ from database to another.

**Table 3.** Nutritional composition of ginger root (USDA, ANSES, DTU).

| | USDA (National Nutrient Database for Standard Reference) | ANSES (Table Ciqual 2016 Composition Nutritionnelle des Aliments) | DTU (Fødevaredatabanken Version 7.01) |
|---|---|---|---|
| Ginger type | Ginger Root, Raw | Ginger, Powder | Ginger Root, Raw |
| Phosphorus (mg/100 g) | 34 | 168 | 34 |
| Magnesium (mg/100 g) | 43 | 214 | 43 |
| Potassium (mg/100 g) | 415 | 1 320 | 415 |
| Manganese (mg/100 g) | - | 33.3 | 0.229 |
| Zinc (mg/100 g) | 0.34 | 3.64 | 0.34 |
| Iron (mg/100 g) | 0.60 | 19.8 | 0.60 |
| Calcium (mg/100 g) | 16 | 114 | 16 |
| Niacin (mg/100 g) | 0.750 | 9.62 | 0.950 |
| Folate (µg/100 g) | 11 | 13 | 11 |

The non-volatile pungent principles responsible of spicy aroma are the gingerols, shogaols, paradols, and zingerone. They represent 4 to 7.5% [27,29] (Figure 2). The principal compound of these is 6-gingerol. In dried ginger, the concentrations of gingerols are reduced whereas the concentrations of shogaols are more abundant, coming from gingerol dehydration [27,32–35].

Volatile oils represent 1–4%. More than 100 compounds are identified. Most of them are terpenoids, mainly sesquiterpenoids (α-zingiberene, zingiberol, β-sesquiphellandrene ... ) and smaller amounts of monoterpenoids (camphene, cineole, geraniol ... ) [34,35].

All the *Z. officinale* samples coming from different origins are genetically indistinguishable. However, their metabolic profiling showed quantitative variation [27] function of the type, the variety but also the agronomic conditions, harvest, drying methods, and storage conditions [35,36].

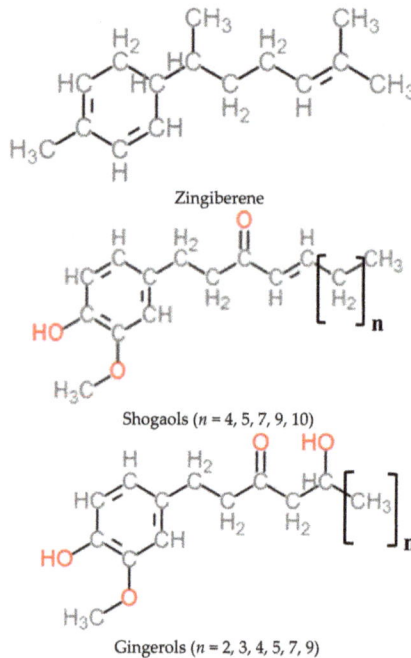

**Figure 2.** Chemical structures of active constituents: zingiberene, shogaols, and gingerols.

## 2.3. Ginger Consumption

In EU countries, the apparent consumption (henceforth referred to simply as 'consumption') of ginger amounted to 58,000 tons in 2014 [22]. A large part of the consumption (around 70–80% of demand) comes from the food processing industry. Dried ginger is especially used in significant amounts in bakery (e.g., gingerbread, cookies) and Asian food products, as well as various drinks (e.g., ginger ale and ginger beer) [22].

Western EU countries accounted for 77% of EU consumption in 2014. The UK (32% of consumed volume), Germany (20%), and the Netherlands (13%) were the largest EU consumers. Concerning the UK, its large ethnic community is an important driver of the national consumption. The volume of consumed ginger in the EU increased by an average of 11% per year between 2010 and 2014, despite the economic crisis. This increase was greatest (over 20% per year) in the Eastern and Northern EU countries [22].

## 2.4. Women's Perception of Ginger

In a multinational survey (Europe, North America, and Australia) including 9113 pregnant women and new mothers, Petersen et al. [37] determined their perception of risks related to medicines, foods or herbal substances, alcohol, tobacco, and thalidomide.

The least harmful products were cranberries and ginger, whereas antidepressants, alcohol, smoking, and thalidomide were rated as the most harmful. Mean risk perception scores for ginger is 1.5 on a 10-point scale (*n* = 8318, Table 4)

**Table 4.** Mean risk perception scores for 13 individual items by geographical region (dark green = 0–2; pale green = 2–4; yellow = 4–6; orange = 6–8; red = 8–10). From Petersen et al. [37].

| | Australia | Eastern Europe | Eastern Europe | Western Europe | North America |
|---|---|---|---|---|---|
| Cranberries | | | | | |
| Ginger | | | | | |
| Eggs | | | | | |
| Paracetamol | | | | | |
| OTC (against nausea) | | | | | |
| Antibiotics | | | | | |
| Swine flu vaccine | | | | | |
| Blue-veined cheese | | | | | |
| Dental X-ray | | | | | |
| Antidepressants | | | | | |
| Alcohol (1st trimester) | | | | | |
| Smoking | | | | | |
| Thalidomide | | | | | |

## 3. Non-Clinical Data

### 3.1. Cytotoxicity

The cytotoxic potential of ginger has been studied for many decades. Various cell lines or in-tissue culture cells were used. Results appeared very different from one study to another (Table 5).

**Table 5.** In vitro cytotoxicity of ginger extracts and main related compounds.

| Ginger, Ginger Extracts, or Related Compounds | Cell Line | IC50 | Reference |
|---|---|---|---|
| Ethanol extract of ginger | CL-6–Calcein assay | 10.95 μg/mL | Plengsuriyakarn et al., 2012 [38] |
| Ethanol extract of ginger | CL-6–Hoechst 33342 assay | 53.13 μg/mL | Plengsuriyakarn et al., 2012 [38] |
| Ethanol extract of ginger | HepG2–Calcein assay | 71.89 μg/mL | Plengsuriyakarn et al., 2012 [38] |
| Ethanol extract of ginger | HepG2–Hoechst 33342 assay | 92.88 μg/mL | Plengsuriyakarn et al., 2012 [38] |
| Ethanol extract of ginger | HepG2 | 358.71 μg/mL | Harliansyah et al., 2007 [39] |
| Ethanol extract of ginger | HRE–Calcein assay | 198.15 μg/mL | Plengsuriyakarn et al., 2012 [38] |
| Ethanol extract of ginger | HRE–Hoechst 33342 assay | 245.91 μg/mL | Plengsuriyakarn et al., 2012 [38] |
| Ethanol extract of ginger | Hamster ovary | 245 μg/mL | Unnikrishnan et al., 1988 [40] |
| Ethanol extract of ginger | Vero cells | 120 μg/mL | Unnikrishnan et al., 1988 [40] |
| Ethanol extract of ginger | Dalton's Lymphoma Ascites | 200 μg/mL | Unnikrishnan et al., 1988 [40] |
| Aqueous extract of ginger | Dalton's Lymphoma Ascites | 420 μg/mL | Unnikrishnan et al., 1988 [40] |
| Aqueous extract of ginger | NRK-52E Cells (MTT test) | No cytotoxicity | Abudayyak et al., 2015 [41] |
| Chloroform extract of ginger | NRK-52E Cells (MTT test) | 9.08 mg/mL | Abudayyak et al., 2015 [41] |
| Methanol ginger extract | MCF7 (Breast) | 75 μg/mL | Zaeoung et al., 2005 [42] |
| Methanol ginger extract | LS174T (Colon) | 80 μg/mL | Zaeoung et al., 2005 [42] |
| Volatile oil of ginger | MCF7 (Breast) | 14.2 μg/mL | Zaeoung et al., 2005 [42] |
| Volatile oil of ginger | LS174T (Colon) | 15.9 μg/mL | Zaeoung et al., 2005 [42] |
| **Related compounds** | | | |
| Diarylheptanoids and gingerol-related compounds | HL-60 | <50 μmol/L | Wei et al., 2005 [43] |
| 6-gingerol | HepG2 | 431.7 μg/mL | Harliansyah et al., 2007 [39] |
| 6-gingerol | MCF7 (Breast) | 31.6 μg/mL | Zaeoung et al., 2005 [42] |
| 6-gingerol | LS174T (Colon) | 30.6 μg/mL | Zaeoung et al., 2005 [42] |
| 6-gingerol | HepG2 | 89.58 μg/mL | Yang et al., 2010 [44] |
| 6-shogaol | MCF7 (Breast) | 6 μg/mL | Zaeoung et al., 2005 [42] |
| 6-shogaol | LS174T (Colon) | 4.2 μg/mL | Zaeoung et al., 2005 [42] |

Wei et al. [43] indicated that diarylheptanoids and gingerol-related compounds were cytotoxic against human promyelocytic leukemia (HL-60) cells (IC50 < 50 µM), while Zaeoung et al. [42] reported that the IC50 of ginger was higher than 39.2 µg/mL against breast (MCF7) and colon (LS174T) cell lines.

In 2008, Kim et al. [45] investigated the cytotoxicity of five compounds of ginger (4-, 6-, 8-, 10-gingerols, and 6-shogaol) on four human tumor cell lines. 6-shogaol showed the most potent cytotoxicity against the four tumor cell lines (IC50 from 1.05 to 1.76 µg/mL) while the others showed moderate activity [45]. Peng et al., for their part, isolated 13 compounds from fresh ginger and studied their cytotoxicity against nine human tumor cell lines. Three of them were identified to be cytotoxic in cell lines tested: 6-shogaol, 10-gingerol, and an enone-diarylheptanoid analog of curcumin [46].

When comparing the IC50 of different chemical constituents to their maximum plasma concentration (see Section 4.1: Pharmacokinetic Data) after an intake of 1.5 or 2 g of ginger, the IC50 value corresponds to concentrations that are not achievable in the human body. IC50 values represent from 7 to ca. 692 times the maximum concentration reached in the plasma (Table 6).

**Table 6.** IC50 values and plasma concentration comparison.

| Ginger Related-Compounds | IC50 (µg/mL) | Plasma Concentration after 1.5 or 2 g of Ginger (µg/mL) | Ratio IC50/Concentration | References |
|---|---|---|---|---|
| 6-gingerol | 15.72 to 431.7 | 1.69 | 9.3 to 255.4 | Zaeoung et al., 2005 [42]; Harliansyah et al., 2007 [39]; Kim et al., 2008 [45]; Zick et al., 2008 [47] |
| 6-shogaol | 1.05 to 6 | 0.0136 to 0.15 | 7 to 441.2 | Zaeoung et al., 2005 [42]; Kim et al., 2008 [45]; Zick et al., 2008 [47]; Yu et al., 2011 [48] |
| 8-gingerol | 8.85 to 12.57 | 0.23 | 38.5 to 54.7 | Kim et al., 2008 [45]; Zick et al., 2008 [47] |
| 10-gingerol | 4.2 to 6.57 | 0.0095 to 0.53 | 7.9 to 691.6 | Kim et al., 2008 [45]; Zick et al., 2008 [47]; Yu et al., 2011 [48] |

Moreover, according to pharmacokinetics data [48] no free 6-, 8-, and 10-gingerols and 6-shogaol were detected in the plasma of all the volunteers 24 h after ginger consumption. Only glucuronide and sulfate metabolites were observed at very low concentration [48]. It is difficult to conclude that ginger is toxic in humans on the basis of these results. The studies presented here cannot be generalized to ginger activity in the body. The potential cytotoxic effects of ginger have to be studied, taking account of the whole plant and its absorption metabolites.

### 3.2. Genotoxicity/Mutagenicity

Ginger showed some mutagenicity in TA 100, TA 1535 [49,50], and T 98 strains [41]. However, ginger's mutagenicity was much lower than established mutagens, such as sodium azide or MNNG [49]. This mutagenicity could be linked to gingerols and shogaols [44,50]. However, in the presence of zingerone, mutagenic activity of gingerols and shogaols was suppressed in a dose-dependent manner (Table 7).

Thus, the observed mutagenic activity of ginger extracts in vitro is the result of the combined action of pro and antimutagenic compounds present in the ginger [50]. However, as for cytotoxic assays, studies presented here cannot be extrapolated to in vivo mutagenicity because of the tested compounds and their concentrations, which are not representative to absorption metabolites and their blood concentration.

Others in vitro and in vivo studies showed antimutagenic properties of ginger or ginger extracts [38,49,51]. When rats fed diets containing 0.5, 1, and 5% of powdered ginger for one month and are exposed to benzopyrene, the mutagenicity in vitro test realized with urine samples showed a reduced number of TA 98 and TA 100 revertants exposed to treated urine at all ginger concentrations compared to control urine [51]. On the basis of ingested dry matter, the ingested daily dose is approximately 0.1, 0.2, and 1 g of ginger by rats, which corresponds to 2.5, 5, and 25 g of ingested powdered ginger in humans, respectively.

Plengsuriyakarn et al., using a crude ethanol extract of ginger, concluded there was an absence of any toxicity at maximum dose of 5 g/kg body weight (BW) using a hamster model [38].

In the in vitro microbial test system, ginger root has mutagenic and antimutagenic properties. In vivo, it appears to be antimutagenic when it is consumed at the human dose.

### 3.3. Acute/Subacute Toxicity (Repeated Dose Toxicity)

Many acute (oral administration of a single dose of a substance in rodents within 24 h) or subacute (oral administration of a test substance in rodents for 28 days) toxicity studies were realized with different rhizome ginger extracts or related compounds. Five studies with ginger powder or ginger extracts (freeze-dried or ethanol extracts) [38,52–55], and one with related compound [56] showed no treatment-related signs of toxicity or mortality in any animals at tested doses. One of them was realized on pregnant rats and showed neither embryotoxic nor teratogenic effects of the tested ginger [55]. The No Observable Adverse Effect Level (NOAEL) was 5000 mg/kg per day whatever the used extract (freeze dried or Ethanol extract) [38,53] whereas the NOAEL of ginger oil is up to 500 mg/kg per day [57]. In four studies [54,58–60], the oral LD50 is far above the highest dose used for human food, dietary supplements or drugs (Table 8).

**Table 7.** Genotoxicity/mutagenicity of ginger and related compounds.

| Test Object | Concentration | Results | References |
|---|---|---|---|
| | In Vitro | | |
| | Ginger | | |
| Ames assay | 5 to 200 µg/plate | Mutagenic on metabolic activation in strains TA 100 and TA 1535 | Nagabhushan et al., 1987 [50] |
| *Salmonella typhimurium* strains TA 100, TA 98, TA 1535, and TA 1538 with and without activation | | | |
| Ames assay | 25 and 50 mg/mL | Mutagenic in both TA, at both concentrations | Soudamini et al., 1995 [49] |
| *S. typhimurium* TA 100 and TA 1535, with and without activation | | | |
| Ames assay | 0.78 to 25 mg/mL | Mutagenic on TA 98 activated cells from 3.13 mg/mL | Abudayyak et al., 2015 [41] |
| *S. typhimurium* TA 100 and TA 98, with and without activation | | | |
| | 6-gingerol | | |
| Comet assay | 0 to 80 µmol/L | DNA strand breaks from 5.89 mg/mL (20 µmol/L) | Yang et al., 2010 [44] |
| HepG2 | | | |
| Ames assay | 5 to 200 µg/plate | Mutagenic on metabolic activation in strains TA 100 and TA 1535 | Nagabhushan et al., 1987 [50] |
| *Salmonella typhimurium* strains TA 100, TA 98, TA 1535, and TA 1538 with and without activation | | | |
| | Shogaol | | |
| Ames assay | 5 to 200 µg/plate | Mutagenic on metabolic activation in strains TA 100 and TA 1535 | Nagabhushan et al., 1987 [50] |
| *Salmonella typhimurium* strains TA 100, TA 98, TA 1535, and TA 1538 with and without activation | | | |
| | Zingerone | | |
| Ames assay | 5 to 200 µg/plate | No mutagenic effect | Nagabhushan et al., 1987 [50] |
| *Salmonella typhimurium* strains TA 100, TA 98, TA 1535, and TA 1538 with and without activation | | | |

Table 8. Acute/subacute toxicity of ginger extracts and the main related compounds.

| | Administration | Species | Oral LD50 | References |
|---|---|---|---|---|
| **Ginger dried rhizome powder** | | | | |
| Ginger powder | per os | Rats | No mortality | Rong et al., 2009 [52] |
| **Ginger dried rhizomes extracts** | | | | |
| Ginger dried rhizomes ethanol extract | intraperitoneal | Mice | $1551 \pm 75$ mg/kg | Ojewole et al., 2006 [58] |
| Ginger dried rhizomes ethanol extract | per os | Hamsters | No mortality | Plengsuriyakarn et al., 2012 [38] |
| Patented standardized ethanol extract of dried rhizomes of ginger (EV.EXT 33) | per os | Pregnant rats | No mortality | Weidner et al., 2001 [55] |
| Freeze-dried ginger powder | per os | Rats | No mortality | Malik et al., 2011 [53] |
| Dry ginger decoction | per os | Rats: Gastric ulcer models | 250 g/kg | Wu et al., 1990 [60] |
| Roasted ginger decoction | per os | Rats: Gastric ulcer models | 170.6 g/kg | Wu et al., 1990 [60] |
| Ginger dried rhizomes ethanol extract | per os | Mice | No mortality at 2.5 g/kg | Anonymous, 2003 [54] |
| Ginger oil | per os | Rabbits | 5 g/kg | Anonymous, 2003 [54] |
| Ginger oil | per os | Rats | No mortality | Jeena et al., 2011 [57] |
| **Related compounds** | | | | |
| **(E)-8 beta,17-epoxylabd-12-ene-15,16-dial (ZT)** | | | | |
| (E)-8 beta,17-epoxylabd-12-ene-15,16-dial (ZT) from ginger | per os (25 mg/kg) | Mice | No mortality | Tanabe et al., 1993 [56] |
| (E)-8 beta,17-epoxylabd-12-ene-15,16-dial (ZT) from ginger | Intra-abdominal (25 mg/kg) | Mice | No mortality | Tanabe et al., 1993 [56] |
| **6-shogaol** | | | | |
| 6-shogaol | intravenous | Mice | 50.9 mg/kg | Suekawa et al., 1984 [59] |
| 6-shogaol | intraperitoneal | Mice | 109.2 mg/kg | Suekawa et al., 1984 [59] |
| 6-shogaol | per os | Mice | 687 mg/kg | Suekawa et al., 1984 [59] |
| **6-gingerol** | | | | |
| 6-gingerol | intravenous | Mice | 25.5 mg/kg | Suekawa et al., 1984 [59] |
| 6-gingerol | intraperitoneal | Mice | 58.1 mg/kg | Suekawa et al., 1984 [59] |
| 6-gingerol | per os | Mice | 250 mg/kg | Suekawa et al., 1984 [59] |

*3.4. Reproductive Toxicity*

One in vivo study was made with a patented ginger extract [55]. Three groups of 22 pregnant female rats from days 6 to 15 of gestation received by gastric intubation of the extract in concentrations of 100, 333, and 1000 mg/kg and were killed on day 21 of gestation. A fourth group received sesame oil as control. Body weight, food, and water intakes were recorded during the study. Standard parameters of reproduction and performance were analyzed. Signs of teratogenic and fetuse toxic effects were determined. No adverse effects or deaths were observed and ginger supplementation was well tolerated. Neither embryotoxic nor teratogenic effects were showed by analysis of fetuses and the authors concluded that the ginger extract administered to pregnant rats during the organogenesis period caused neither maternal nor developmental toxicity at doses of up to 1 g/kg body weight per day [55].

By contrast, some ginger adverse events have been reported on pregnant rats [61]. Pregnant Sprague–Dawley rats received ginger tea (15, 20, or 50 g/L) on day 6 of gestation onwards until day 15 of gestation. They were sacrificed at day 20. No maternal toxicity was observed, but embryonic loss was increased in the treatment group compared to control. No gross morphologic malformations were seen but fetuses exposed to ginger tea were significantly heavier than control and had more advanced skeletal development. This effect was greater in female fetuses and not correlated with increased placental size. Thus, results of this study suggest that in utero exposure to ginger tea leads to an increased early embryo loss with increased growth in surviving fetuses [61].

In 2009, Sukandar et al. studied the effect of ethanol extracts of ginger and noni fruit in pregnant rats. The blend was administered per os at three different doses on days 6–15 of gestation (50 and 50, 500 and 500, 1000 and 1000 mg/kg BW). Sacrifices were realized on day 19 of gestation to observe live fetus, resorption, and growfail fetus. No malformation was found. The combination, did not cause any fetus resorption, growth failure, malformation in organs nor in the skeleton, whatever the dose. The highest dose (500 and 500 and 1000 and 1000 mg/kg BW) caused liver color change of 7.3% and 8.3% of rat fetuses, respectively [62].

In a double-blind randomized cross-over clinical trial, ginger per os, 250 mg, four times daily did not generate teratogenic aberrations in newborns and Apgar scores of these babies were 9 to 10 after five minutes [63].

Reproductive and developmental toxicity has been investigated in three studies in rats [55,61,62] and one double-blind randomized cross-over clinical trial [63]. All these studies showed no teratogenic aberrations. Extensive data do not suggest any major concerns with respect to reproductive and developmental safety of ginger root.

## 4. General Population Safety Data

Many reviews have been published on ginger effects and few minor adverse events have been reported with the use of ginger in humans [34,64–67].

A systematic review by Betz et al. [65] including 15 randomized studies with safety data showed that, among the 777 patients included, 3.3% reported slight side effects that did not require treatments like mild gastrointestinal symptoms and sleepiness.

In a clinical trial involving 12 healthy volunteers who consumed 400 mg of ginger three times per day for two weeks, one subject reported mild diarrhea during the first two days of treatment. Authors explained that ginger could cause heartburn and as a gastric irritant with doses higher than 6 g. Moreover, its inhalation could produce an IGE-mediated allergy [64].

An assessment report on *Zingiber officinale* Roscoe made by the Committee on Herbal Medicinal Products (HMPC) in 2012 summarized the published clinical studies with safety data indexed by PubMed through June 2010 [30].

Adverse effects resulting from the intake of ginger root observed in clinical studies occur with low frequency, low intensity, and are mainly gastrointestinal. No severe events have been reported. An interaction between ginger and warfarin has been showed in some case reports however the latter are unconvincing. Moreover, one randomized study in healthy volunteers failed to demonstrate any warfarin interaction. No sufficient evidence are available to suggest induction or inhibition of CYP-enzymes by ginger or its constituents [30]. Consequently, the committee concluded that the benefit/risk balance is in favor of the oral use of ginger extract and complies with the criteria for well-established use in the prevention of nausea and vomiting in motion sickness [30]. However, ginger's effect on platelet aggregation cannot be confidently dismissed and future clinical trials are needed to further investigate this area, particularly in risk population [68].

More recently, Wang et al. [69] assessed daily ginger consumption in adults, explored its correlation with chronic diseases and analyzed further how different levels of ginger intake may affect the prevalence of chronic diseases. A total of 4628 participants (1823 men and 2805 women), aged from 18 to 77 years old, completed face-to-face dietary and health questionnaires. Daily ginger consumption was associated with a decreased risk for hypertension ((OR), 0.92; 95% confidence interval (CI), 0.86–0.98) and chronic heart disease (CHD) (OR, 0.87; 95% CI, 0.78–0.96) in adults $\geq$18 years old. Differences were also observed in adults $\geq$40 years old: hypertension (OR, 0.92; 95% CI, 0.87–0.99), CHD (OR, 0.87; 95% CI, 0.78–0.97). However, after 60 years old, no association was seen for hypertension, but there was still a difference regarding CHD (OR, 0.84; 95% CI, 0.73–0.96). Again, the probability of illness (hypertension or CHD) decreased when the dosage of daily ginger intake increased. No adverse events were reported [69].

*4.1. Pharmacokinetic Data*

In 2008, Zick et al. [47] realized a single dose pharmacokinetic escalation study of the ginger constituents 6-gingerol, 8-gingerol, 10-gingerol, and 6-shogaol. 27 healthy volunteers were recruited, three participants per dose except for the highest doses for which 6 and 9 participants were included. Administrated doses of ginger extract were 100 mg, 250 mg, 500 mg, 1.0 g, 1.5 g, and 2.0 g. The dry extract of ginger root used in the study was standardized to 6% of gingerols. A dose of 250 mg correspond to 15 mg of total gingerols, with 5.38 mg 6-gingerol, 1.28 mg 8-gingerol, 4.19 mg 10-gingerol, and 0.92 mg 6-shogaol. Blood was sampled at different times after ginger intake and analysis showed that no free 6-gingerol, 8-gingerol, 10-gingerol, or 6-shogaol was detected and no conjugate metabolite were detected below the 1 g ginger extract dose, except for 6-gingerol. However, as of this dose, all metabolites were quickly absorbed and detected as glucuronide and sulfate conjugates. Conjugate metabolite appeared 30 min after 2 g dose intake, with their $T_{max}$ between 45 and 120 min and their elimination half-lives between 75 to 120 min. The maximum concentrations were 1.69 µg/mL for 6-gingerol, 0.23 µg/mL for 8-gingerol, 0.53 µg/mL for 10-gingerol, and 0.15 µg/mL for 6-shogaol at either the 1.5 g or 2.0 g dose. No pharmacokinetic model was able to be constructed due to the low levels of ginger constituent absorption, and the pharmacokinetic parameters were based on a non-compartment analysis with an elimination half-life only presented for the 2.0 g dose [47].

Another study made by Yu [48] in 2011 with 12 volunteers receiving 2 g of ginger extract per os for 24 days showed that no 6-, 8-, and 10-gingerols, and 6-shogaol under free form were detected in the plasma of all the participants 24 h after the last dosing. Concentrations of 6-gingerol glucuronide (from 5.43 to 13.6 ng/mL), 6-gingerol sulfate (from 6.19 to 7.29 ng/mL), and 10-gingerol glucuronide (from 6.96 to 9.33 ng/mL) were low whereas levels of other conjugate metabolites were not detectable in all the volunteers. Maximum concentrations determined for 10-gingerol and 6-shogaol were 9.5 $\pm$ 2.2 ng/mL (0.027 $\pm$ 0.006 µM) and 13.6 $\pm$ 6.9 ng/mL (0.049 $\pm$ 0.025 µM), whereas the IC50s of 10-gingerol and 6-shogaol were 12 and 8 µM, or 4.2 and 2.2 mg/mL, respectively. The half-lives of all metabolites are between 1 and 3h [48].

Table 6 reviews the pharmacokinetic data and IC50 values of some ginger metabolites absorbed and detected in plasma, such as 6-, 8-, 10- gingerols, and 6-shogaol. As mentioned before, IC50 values represents between 7 to 692 times the maximum concentrations detected in plasma.

## 5. Pregnant Women Data

*5.1. Safety Data*

We analyzed 14 randomized clinical studies from 1991 [63] to 2017 [70] and three prospective studies [71–73]. In randomized clinical studies, a total of 1 331 pregnant women have been studied, including 617 who consumed ginger (Tables 9 and 10). Paritakul et al. studied 63 postpartum women (33 for the placebo group and 30 for ginger group). Studies included pregnant women at less than 20 weeks of gestation and only for one greater than or equal to 37 weeks of gestation [74].

Different types of ginger were used, like fresh ginger, ginger powder, ginger extract, or ginger essence. The ginger dose ranges studied were from 500 mg/d to 2.5 g/d. The recommended daily doses were: 3 × 650 mg, 4 × 250 mg, 2 × 500 mg, 4 × 125 mg, 3 × 350 mg, 3 × 250 mg, and five biscuits. The majority of the studies were conducted over four days, but three studies chose seven days [75–77], one study was conducted for two weeks [78], and one study was conducted for three weeks [79]. Clinical studies were realized in Asia (Thailand), Europe (Denmark), Oceania (Australia), and the Middle East (Iran). The majority of studies (12/14) lasted between four and seven days, which is consistent with the symptomatology of pregnant women with regard to nausea and vomiting. Ten studies, all randomized versus placebo or positive control (vitamin B6 or dimenhydrinate), have highlighted various side effects.

**Table 9.** Randomized double-blind trials investigating the effectiveness and safety of ginger for pregnancy.

| Objective | Population | Number | Treatment | Ginger Definition | Duration | Results | Adverse Events | Reference |
|---|---|---|---|---|---|---|---|---|
| To determine the effectiveness of ginger for the treatment of NVP | Women NVP < 17 weeks of gestation | $n = 67$ 35 placebo 32 ginger | 1000 mg/day (4 × 250 mg) of ginger (powder capsules) vs. placebo | Fresh ginger root | 4 days + follow-up visit 7 days later | Significant decrease of nausea in the ginger group vs. placebo group ($p = 0.014$) Significant decrease of vomiting in the ginger group vs. placebo group ($p < 0.001$) Follow-up visits:Significant symptom improvement in the ginger group vs. placebo group ($p < 0.001$) | Headache: - 5 in the placebo group - 6 in the ginger group. Ginger group: - 1 abdominal discomfort - 1 heartburn - 1 diarrhoea for one day **Side effects reported as minor.** Spontaneous abortions: - 3 in the placebo group - 1 in the ginger group Term delivery: - 91.4% in the placebo group - 96.9% in the ginger group Cesarean deliveries: - 4 in the placebo group - 6 in the ginger group **No infants had any congenital anomalies recognized. No significant adverse effect of ginger on pregnancy outcome was reported.** | Vutyavanich et al., 2001 [80] |
| To compare the effectiveness of ginger and vitamin B6 for treatment of NVP | Women NVP < 16 weeks of gestation | $n = 123$ 62 vit. B6 61 ginger | 1950 mg/day of ginger (3 × 650 mg) or 75 mg/day of vitamin B6 (3 × 25 mg) | Fresh ginger root | 4 days | Nausea/vomiting: Improvement of nausea vomiting scores in both group from baseline The average score change in the ginger group was better than that of vitamin B6 group ($p < 0.05$) | Side effects: - 16 in ginger group - 15 in B6 group (NS) Heartburn: - 8 in ginger group - 2 in B6 group Sedation: - 7 in ginger group - 11 in B6 group Arrhythmia: - 1 in ginger group Headache: - 2 in B6 group **Side effects were reported to be minor** | Chittumma et al., 2007 [81] |
| To determine the effectiveness of ginger for the treatment of NVP | Women NVP < 20 weeks of gestation | $n = 67$ 35 placebo 32 ginger | 1000 mg/day (4 × 250 mg) of ginger (powder capsules) | Ginger root powder (Zintoma, Goldaroo Company, Tehran, Iran) | 4 days | Nausea: Significant improvement in 84% of ginger users vs. 56% in the control group ($p < 0.05$) Vomiting: Significant improvement in the treated group vs. the control group ($p < 0.05$) | No complication during the treatment period was reported **Reported as a safe remedy to improve the nausea and vomiting of pregnancy.** | Ozgoli et al., 2009 [82] |

**Table 9.** *Cont.*

| Objective | Population | Number | Treatment | Ginger Definition | Duration | Results | Adverse Events | Reference |
|---|---|---|---|---|---|---|---|---|
| To compare the effectiveness of ginger and vitamin B6 for the treatment of NVP | Women with nausea with or without vomiting < 17 weeks of gestation | $n = 69$ 34 vit. B6 35 ginger | 1000 mg/day (2 × 500 mg) of ginger (powder capsules) or 40 mg/day of vitamin B6 (2 × 20 mg) | Fresh ginger root | 4 days + follow-up visit 7 days later | Nausea score: Better score in the ginger group vs. the vitamin group ($p = 0.024$) Vomiting episodes: No significant difference between the groups Follow-up visits: 82.8% reported an improvement in the ginger group vs. 67.6% in the vitamin group ($p = 0.52$) | Spontaneous abortions: - 2 in the ginger group - 1 in the B6 group ($p > 0.05$) Term birth: - 82.9% in the ginger group - 82.4% in the B6 group Caesarean deliveries: - 4 in the ginger group - 6 in the B6 group ($p > 0.05$) **No babies had any congenital anomalies All were discharged in good condition No adverse effects of ginger on pregnancy outcome were reported** | Ensiyeh et al., 2009 [83] |
| To compare the effects of ginger on nausea and vomiting caused by pregnancy and compares it with metoclopramide medicine | Women NVP < 20 weeks of gestation | $n = 102$ 34 placebo 34 metoclopramide 34 ginger | 600 mg/day (3 × 200 mg) ginger;30 mg/day (3 × 10 mg) metoclopramide;600 mg/day (3 × 200 mg) placebo | Ginger essence | 5 days | Intensity of nausea: Significant difference in the two supplemented groups (ginger or metoclopramide) vs. placebo ($p < 0.05$) Not statistically significant between treated groups | - | Mohammadbeigi et al., 2011 [84] |
| To determine if ginger syrup mixed in water is an effective remedy for the relief of NVP | Women with nausea with or without vomiting first trimester | $n = 23$ 10 placebo 13 ginger | 1 g/day (4 × 250 mg) ginger (tablespoon) vs placebo | Ginger including 1 mg pungent compounds from ginger rhizome juice, 1 mg of 20% pungent compounds and 5% zingiberene coming from $CO_2$ supercritical extract of ginger rhizome | 2 weeks | Nausea: 77% improvement in ginger group vs. 20% in placebo group Vomiting: 67% in ginger group stop vomiting at day 6 vs 20% in placebo group | Delivered viable infant at term without major complications **Safe option in the treatment of NVP** | Keating et al., 2002 [78] |

**Table 9.** *Cont.*

| Objective | Population | Number | Treatment | Ginger Definition | Duration | Results | Adverse Events | Reference |
|---|---|---|---|---|---|---|---|---|
| To investigate the effect of a ginger extract (EV.EXT35) on the symptoms of morning sickness. | Women with morning sickness < 20 weeks of gestation | *n* = 9951 placebo48 ginger | 500 mg/day (4 × 125 mg) eq. 6 g/day ginger vs. placebo | EV.EXT35: ginger extract (125 mg eq to 1.5 g of dried ginger) | 4 days | Nausea experience score: - except for day 3, the difference parameter for each day post-baseline, was significantly less than zero Vomiting; - no significant difference between the groups | Spontaneous abortion: - 3 in the ginger group - 1 in the placebo group Intolerance of the treatment: - 4 in the ginger group Worsening of treatment requiring further medical assistance: - 1 in the ginger group - 2 in the placebo group Allergic reaction to treatment: - 1 ginger group **No apparent increased risk of fetal abnormalities or low birth weight.** | Willetts et al., 2003 [85] |
| To estimate whether the use of ginger to treat nausea or vomiting in pregnancy is equivalent to pyridoxine hydrochloride (vitamin B6) | Women NVP Between 8 and 16 weeks of gestation | *n* = 235 115 vit. B6 120 ginger | 1.05 g/day ginger (3 × 350 mg) vs. 75 mg/day vitamin B6 (3 × 25 mg) | – | 3 weeks | 53% reported an improvement taking ginger, and 55% reported an improvement with vitamin B6 Ginger was equivalent to vitamin B6 for improving nausea, dry retching, and vomiting | Belching: - ginger (9%) vs. B6 (0%) (*p* < 0.05) Dry retching after swallowing: - ginger (52%) vs. B6 (56%) Vomiting after ingestion: - ginger (2%) vs. B6 (1%) Burning sensation: - ginger (2%) vs. B6 (2%) Pregnancy outcome: - 272 (93%) gave birth to 278 infants - 12 women with spontaneous abortion (first or second trimester) - 3 women with stillbirth - No differences were found between study groups - 9 babies born with congenital abnormality (3 ginger, 6 B6) (NS) - 6 cases of urogenital disorders - 2 cases of minor gastrointestinal abnormalities - 1 case of a minor congenital heart defect | Smith et al., 2004 [79] |

Table 9. *Cont.*

| Objective | Population | Number | Treatment | Ginger Definition | Duration | Results | Adverse Events | Reference |
|---|---|---|---|---|---|---|---|---|
| To examine the evidence for the safety and effectiveness of ginger for NVP | Women NVP Between 7 and 17 weeks of gestation | $n = 62$ 30 placebo 32 ginger | 5 biscuits/day (2.5 g of ginger) vs. placebo | - | 4 days + follow-up visit 7 days later | Nausea scores: - significantly greater in the ginger group vs. in placebo group ($p = 0.01$) Vomiting episodes: - no significant difference ($p = 0.243$) No vomiting after 4 days: -34% in ginger group vs. 18% in placebo group Follow-up visits: -87.5% ginger group reported improvement vs. 70% in placebo group ($p = 0.043$) | In ginger group: -1 dizziness -1 heartburn **The side effects reported as minor. No abnormal pregnancy and delivery outcome occurred. No infants had any congenital abnormalities recognized. All discharged in good condition** | Basirat et al., 2009 [86] |
| To compare the breast milk volume during the early postpartum period between women receiving dried ginger capsules with those receiving placebo | Women $\geq 37$ weeks gestation | $n = 63$ 33 placebo 30 ginger | 1 g of ginger (500 mg × 2) vs. placebo | Dried ginger powder | 7 days | Breast milk volume: - day 3 ginger group has higher milk volume than the placebo group ($p < 0.01$) - day 7, the ginger group does not differ from the placebo group Prolactin levels is similar in both groups | No notable side effects | Paritakul et al., 2016 [74] |
| To study the efficacy of ginger and dimenhydrinate in the treatment of NVP | Women NVP < 16 weeks of gestation | $n = 170$ 85 dimenhydrinate 85 ginger | 1 g (500 mg × 2) of ginger vs. 100 mg (50 mg × 2) of dimenhydrinate | - | 1 week | Nausea: - the mean score in day 1-7 decreased in both groups - daily mean scores between both groups were not statistically different Vomiting: - frequency of vomiting times in day 1-7 decreased in both groups - daily mean vomiting times in the dimenhydrinate group in day 1-2 were less than the ginger group ($p < 0.05$) - after day 3-7 post treatment, the daily mean vomiting times in both groups were not statistically different | Drowsiness: -5/85 in the ginger group vs. 66/85 in dimenhydrinate group ($p < 0.01$) Heart burn: -13/85 in the ginger group vs. 9/85 in dimenhydrinate group ($p = 0.403$) **No other adverse effect was reported in both groups** | Pongrojpaw et al., 2007 [75] |

Table 9. Cont.

| Objective | Population | Number | Treatment | Ginger Definition | Duration | Results | Adverse Events | Reference |
|---|---|---|---|---|---|---|---|---|
| To study the efficacy of ginger and placebo in hyperemesis gravidarum. | Women hyperemesis gravidarum < 20 weeks of gestation | $n = 27$<br>13 lactose<br>14 ginger | 1 g (250 mg × 4) of ginger vs. placebo | Powdered root | 2 × 4 days<br>2 days washout | The preference:<br>- ginger treatment period was statistically significant ($p = 0.003$)<br>Relief of the hyperemesis symptoms:<br>- significantly greater in ginger group vs. placebo ($p = 0.035$) | One spontaneous abortion, which was not a suspicious high rate of fetal wastage in early pregnancy<br>No side effects were observed<br>All infants were without deformities and discharged in good condition. | Fischer-Rasmussen et al., 1991 [63] |
| To compare the effectiveness of ginger and acupressure in the treatment of NVP | Women NVP < 16 weeks of gestation | $n = 143$<br>45 control<br>48 acupressure | 750 mg (250 mg × 3) of ginger vs. acupressure | - | 7 days<br>- 3 with no intervention<br>- 4 with treatment | Rhodes index scores:<br>- better in the ginger group vs. acupressure and control ($p < 0.001$)<br>- reduced 49% in ginger group and 29% in acupressure group.<br>- increased up to 0.06% in control group<br>Post hoc test showed significant differences in vomiting, nausea, retching, and total scores between the groups except for vomiting score between acupressure and control groups and for retching score between acupressure and ginger groups | 1 case of heartburn with ginger capsules | Saberi et al., 2013 [76] |
| To determine the effect of ginger to relieve NVP | Women NVP < 16 weeks of gestation | $n = 106$<br>36 placebo<br>33 control<br>37 ginger | 750 mg (250 mg × 3) of ginger vs. acupressure | - | 7 days<br>- 3 with no intervention<br>- 4 with treatment | Rhodes index scores:<br>- greater in the ginger group vs. placebo and control ($p < 0.001$)<br>- reduced 48% in ginger group, 13% in placebo group and -10% in control group<br>Post hoc test showed significant difference between the groups in reduction of vomiting, nausea, retching and total Rhodes Index scores ($p < 0.001$) | 1 case of heartburn with ginger capsules | Saberi et al., 2014 [77] |
| To compare the effects of ginger, pyridoxine (vitamin B6), and placebo for the treatment of NVP | Women between 6 and 16 weeks of pregnancy; mild and moderate NVP | $n = 77$<br>23 placebo<br>26 vit. B6<br>28 ginger | 1 g (500 mg × 2) ginger capsules<br>80 mg (40 mg × 2) Vit. B6 capsules | - | 4 days | Rhodes index scores:<br>- ginger > placebo ($p = 0.039$)<br>- Vit. B6 > placebo ($p = 0.007$)<br>- ginger = Vit. B6 ($p = 0.128$)<br>Ginger was more effective for<br>- nausea intensity<br>- nausea distress<br>- distress of vomiting | Ginger is effective and safe | Sharifzadeh et al., 2017 [70] |

**Table 10.** Prospective studies investigating the effectiveness and safety of ginger for pregnancy.

| Study Type | Objective | Population | Number | Treatment | Ginger Definition | Duration | Results | Adverse Events | Reference |
|---|---|---|---|---|---|---|---|---|---|
| Prospective cohort study (Korean Motherisk Program) | To determine if ginger exposure during pregnancy would increase the risk of adverse fetal and neonatal outcomes | Women Cough and cold preparations (49.7%) Functional gastrointestinal disorders (37.7%) 3 days to 20.7 weeks | 159 306 (control group) | Median dose: 470 mg/day Maximum dose: 7.2 g/day | Dried ginger | Median length: 2 days | | Spontaneous abortion: -7 in the ginger group vs. 17 in the control group (NS) Stillbirths: -2.7% in the ginger group vs. 0.3% in the control group (NS) NICU admission: -4.7% in the ginger group vs. 1.7% in the control group (NS) The admission rate was marginally different between cases and controls however, no difference was observed when admission rate was compared with that reported in 2012 at the hospital. **They concluded that dried ginger is not a major human teratogen** | Choi et al., 2014 [73] |
| The Norwegian Mother and Child Cohort study | To evaluate the safety of ginger use during pregnancy on congenital malformations | Women NVP first trimester | n = 68 522 -1020 (use ginger) -466 (use ginger during 1st trimester) | | | | | Not increase the risk of malformations No significant associations between ginger use risk of: - Stillbirth - Perinatal death - Low birth weight - Preterm birth - Low Apgar score | Heitmann et al., 2013 [72] |
| Prospective cohort study (The Motherisk Program) | The primary objective: - to examine the safety The secondary objective: - to examine the effectiveness of ginger for NVP | Women first trimester | 187 (ginger group) 187 (comparison group) | Various types of ginger: - Capsules (49%) - Ginger tea - Fresh ginger - Pickled ginger - Ginger cookies - Ginger candy - Inhaled powdered ginger - Ginger crystals - Sugared ginger | | Minimum 3 days | Ginger effectiveness scores (overall) on 66 women -3.6 ± 2.4 (SD) Scale range of effectiveness: -0 => 29 (43.9%) -2-4 => 19 (28.8%) -5-7 => 13 (19.7%) -8-10 => 6 (7.6%) 0 = No effect; 1-4 = mild effect; 5-7 = moderate effect; 8-10 = best effect. | -3 spontaneous abortions -2 stillbirths -1 therapeutic abortion (Down syndrome) 3 major malformations in the ginger group: - ventricular septal defect - lung abnormality - kidney abnormality (pelviectasis) -1 idiopathic central precocious puberty at 2 years old No significant differences between the groups in terms of live births, spontaneous abortions, stillbirths, therapeutic abortions, birth weight, or gestational age. Eight sets of twins in the ginger group. More babies < 2.5 Kg in the control group | Portnoi et al., 2003 [71] |

The various adverse effects founded in RCTs are listed in Table 11. There are no common side effects from one study to another with the exception of heartburn and spontaneous abortion, which are quoted six and five times, respectively.

**Table 11.** List of adverse effects identified in the different clinical studies.

| Adverse Effects Identified (No Significance) | References |
|---|---|
| Headache, abdominal discomfort, diarrhea, heartburn, spontaneous abortion | Vutyavanich et al., 2001 [80] |
| Sedation, arrhythmia, heartburn | Chitumma et al., 2007 [81] |
| Intolerance, allergic reaction, medical assistance, spontaneous abortion | Willets et al., 2003 [85] |
| Dry retching, vomiting, burning sensation, belching, spontaneous abortion | Smith et al., 2004 [79] |
| Dizziness, heartburn | Basirat et al., 2009 [86] |
| Drowsiness, heartburn | Pongrojpaw et al., 2007 [75] |
| Spontaneous abortion | Ensiyehh et al., 2009 [83] |
| Spontaneous abortion | Fisher Rasmussen et al., 1991 [63] |
| Heartburn | Saberie et al., 2013 [76] |
| Heartburn | Saberie et al., 2014 [77] |

This diversity of side effects can be explained by changes in several parameters depending on the studies. The most important point is that ginger quality is never specified and there are many variations on the type of ginger, recommended daily doses, and the duration of treatment.

According to Boltman-Binkowski [87], ginger does not increase spontaneous abortion compared to the control group (relative risk with 95% CI = 0.80 (0.21, 2.99)) and does not increase the rate of stillbirth (relative risk with 95% CI = 0.64 (0.03, 13.59)) and congenital abnormalities (relative risk with 95% CI = 2.11 (0.07, 65.87)) compared to the general population.

In Saberi's studies made in 2013 and 2014 [76,77], a total of 249 women participated to RCTs, 87 of whom are treated with ginger, and only two cases of heartburn were detected. Babies whose mothers were exposed to ginger during Saberi's study did not appear to be at an increased risk of fetal abnormalities or low birth weight. All these side effects were reported by subjects as minor and did not preclude them from taking their prescribed medication [76,77].

Three prospective studies [71–73] focus on the exposure of pregnant women to ginger and the possible side effects that would result from it. In 69 361 women, 812 pregnant women were studied because they took ginger during first trimester. Choi et al. [73] noted seven cases of spontaneous abortions in the ginger group and 17 cases in the control group (OR: 0.8; 95% CI: 0.3–1.9; $p$ = 0.59). At first reading, rates of stillbirths and babies that required attention at the neonatal intensive care unit (NICU) were marginally superior in the ginger group vs. the control group. However, no difference was observed when the NICU admission rate was compared with what was reported from the hospital and Women's Healthcare Center in 2012.

Heitmann et al. [72] found that ginger use during pregnancy at any time did not increase the risk of malformations (4.7% in the no exposure to ginger group vs. 4,1% in those exposed to ginger during the first trimester group, adjusted OR (95 % CI) 0.8 (0.5–1.4)). Moreover, ginger use during pregnancy is not significantly associated to the risk of stillbirth/perinatal death, low birth weight, preterm birth, or low Apgar score. Equivalent results were found by Portnoi et al. [71], with no differences between the groups in terms of live births, spontaneous abortions, stillbirths, therapeutic abortions, birth weight, or gestational age. One exception: there were more infants who weighed less than 2500 g in the control group. Unfortunately, Heitmann et al. [72] specified neither the duration, nor the dose, nor the type of ginger used. Several types of ginger were consumed by women in Portnoi's study [71] and Choi et al. [73] did not give any details except that it is a dried ginger.

Three studies [70,78,82], used ginger in syrup and powder form, respectively, on 14, 28, and 32 pregnant women during two weeks or four days. None of them had safety outcomes, but they concluded that ginger is a safe option in early pregnancy.

Term birth parameter is mentioned in three others studies [79,80,83]. In two of them, there was no significant difference between ginger groups vs. other groups (placebo or B6) [80,83]. Conversely,

Smith, et al., showed that term birth in the ginger group was significantly better than placebo (93% vs. 98% for placebo and of ginger groups, respectively, *p* = 0.03) [79].

A randomized, double-blind controlled trial was conducted by Paritakul et al. [74] in 2016 on women for seven days postpartum. Sixty-three women (30 ginger group, 33 placebo group) received 500 mg of dried ginger powder in capsules twice a day and no notable side effects are mentioned.

The 14 randomized and three prospective clinical studies have observed more than 70,000 pregnant women, including more than 1500 who have consumed ginger during their pregnancy. Studies ran through over more than 25 years. Doses, durations, types of ginger, and countries vary from one study to another, thus addressing the different possibilities of current consumption. The use of ginger during pregnancy do not present a risk for the mother or her future baby. All studies conclude the safety of ginger. We can, however, note a side effect inherent to the composition of ginger—heartburn—which must be monitored in sensitive persons.

According to McLay [88], ginger could potentially cause interactions with concurrent prescription medicines: one major interaction with nifedipine, and three moderate interactions with metformin, insulin, and aspirin. In France, these drugs—except insulin—are not recommended for use or are contraindicated in pregnant women.

## 5.2. Efficacy

The mechanisms of action underlying ginger's efficiency in reducing NVP has been investigated [89], dual antiemetic action have been highlighted: (i) gingerols and shogaols act as antagonists of cholinergic M3 and serotonin 5-HT3 receptors of the central nervous system; (ii) ginger's constituents improve the gastric tonus, motility, and emptying due to peripheral anticholinenergic and antiserotonergic actions. However, these data need further investigations to elucidate and confirm those preliminary findings.

Concerning the efficacy of ginger on pregnant women, Mohammadbeigi's study is added to the previous studies mentioned in Section 5.1 [84]. The total number of pregnant women studied in randomized clinical studies was 1433, with 651 pregnant women who consumed ginger. The doses used and the duration of treatment vary according to the studies.

To determine the effectiveness of ginger on NVP, its effect was compared to control [77] or placebo [77,78,80,84–86] groups, but also vitamin B6 [70,79,81,83], drugs like dimenhydrinate [75], metoclopramide [84], or acupuncture [76].

A minimum of 1 g of fresh ginger root per day during at least four days significantly decreased nausea and vomiting and improved significantly symptoms vs. placebo or vitamin B6 [79–81,83].

Willets, et al., [85] investigated in a randomized double-blind placebo-controlled study, the effect of a ginger extract (EV.EXT35) on morning sickness symptoms in 120 pregnant women. 125 mg of ginger extract was equivalent to 1.5 g of dried ginger. After four days, the nausea experience score was significantly less than zero (except for day 3). Concerning vomiting symptoms, there was no significant difference between ginger extract and placebo groups

Keating et al. [78] demonstrated that 1 g/d (4 × 250 mg) of ginger syrup for a two-week period revealed that 67% of women in the ginger group stopped vomiting at day 6 vs. 20% in the placebo group. Additionally, 77% of women in the ginger group had a four-point improvement on the nausea scale at day 9 vs. 20% in the placebo group.

Mohammadbeigi et al. [84] used 600 mg/d (3 × 200 mg) of ginger essence for five days vs. 30 mg/d (3 × 10 mg) of metoclopramide and 600 mg/d (3 × 200 mg) of placebo. Significant decreases in the severity of nausea and vomiting, as well as the Rhodes Index for ginger and metoclopramide groups vs. the placebo group, was shown. There was also no difference between both groups.

Basirat et al. [86] realized a randomized double-blind clinical trial on 62 pregnant women. Thirty-two women took five biscuits daily for four days. Each biscuit contained 0.5 g of ginger. The average change in nausea scores in the ginger group was significantly better (*p* = 0.01) than that in the placebo group. The average change in the number of vomiting episodes was not significantly different between the ginger group and the placebo group. However, after four days of treatment, the proportion of women who had no vomiting in the ginger group (11/32 patients) was greater than that in the placebo group (6/30).

Regardless of the dose, duration, or type of ginger, there is a significant impact on nausea and there are no safety concerns. In 4/15 studies either ginger had no effect on vomiting or there was no significant difference with placebo or vitamin B6.

According to Ding et al. [90], all various forms of ginger studied were a safe and effective treatment for NVP when compared to placebo and vitamin B6. Two meta-analyses made in 2014 [67,91] concluded also that ginger could be considered as an effective and harmless alternative option for women suffering from the symptoms of NVP and that 1 g/day for a duration of at least four days is better than placebo in improving NVP.

## 6. Conclusions

NVP affect 7 in 10 pregnant women and have a deep impact on the quality of life. Recently, increasing concerns have been pointed out regarding the safety of traditional antiemetic drugs (metoclopramide, domperidone, etc.). Thus, considering natural options, such as ginger, with a favorable risk/benefit ratio and a good level of evidence, is now part of several practice guidelines [92].

Many food supplements with ginger powder or ginger extracts are used to decrease symptoms of nausea and vomiting associated with pregnancy. This effect is supported by a European claim (ID 2172) that it *"Helps to support the digestion/contributes to the normal function of intestinal tract/contributes to physical well-being/contributes to the normal functioning of the stomach in case of early pregnancy"* provided the product contains the equivalent of 0.5 to 2 g of root per day.

Beyond the quantity of ginger needed to be effective, the quality of the ginger is important for the safety aspect. For example, in France, it is mandatory to monitor the concentration of methyleugenol because of its potential toxicity [5]. It is important to focus special attention on pungent components of ginger powder, including the main components, 6-gingerol, 8-gingerol, and 10-gingerol [89]. It is also important to highlight the wide variability in 6-gingerol, 6-shogaol, 8-gingerol, and 10-gingerol composition from one food supplement to another. The 6-gingerol concentration ranged from 0.0 to 9.43 mg/g, 6-shogaol ranged from 0.16 to 2.18 mg/g, 8-gingerol ranged from 0.0 to 1.1 mg/g, and 10-gingerol ranged from 0.0 to 1.40 mg/g [93]. Variations could be due to sourcing, method and period of harvest, storage, and processing methods [93]. The main critical control points to assure safety and quality of ginger used are chemical constituents (actives like gingerols, or potentially toxics like methyleugenol), contaminants (microbiology, pesticides, heavy metals, residual solvents), and adulteration risks. In view of the sensitivity of this raw material, control and qualification procedures appear mandatory: supply chain transparency, traceability, management of material safety, and quality standards are keys to assure consumer safety. Processes have to be in place to approve the suppliers' production sites and the relationship between buyer and supplier is critical to support any adulteration prevention effort.

With regard to toxicity associated with ginger chemical constituents, such as gingerols, cytotoxic or mutagenic in vitro studies already performed are not representative and difficult to extrapolate to humans. Potential toxic effect of ginger have to be studied taking account of the whole plant and its absorption metabolites in humans (glucuronides and sulfates of gingerols, for example), not plant metabolites. Recently, one in vitro study showed that free forms are more cytotoxic compared to the glucuronide conjugates [94].

All the in vivo results do not suggest any major concerns with respect to reproductive and developmental safety of ginger root. At least, no associations were found between the use of ginger and malformations in humans. This finding is reassuring and supports previous findings. Moreover, according to the data, the use of ginger during pregnancy does not increase the risk for any of the following pregnancy outcomes: stillbirth/perinatal death, low birth weight, preterm birth, and low Apgar score [72].

It is interesting to note that one member of the Committee on Herbal Medicinal Products (HMPC) did not agree with HMPC's opinion on *Zingiber officinale* Roscoe rhizome [30,95]. He reported that ginger is used in food without any restrictions. Additionally, there are results of tests on reproductive toxicity and results of clinical trials including pregnant women published which support safe use during pregnancy. Therefore, the restriction for use during pregnancy is not justified. Moreover the results of the clinical trials for pregnancy-induced vomiting are such a quality that well-established use could be established for ginger rhizomes [92].

Finally, a recent consensus [96] was published on a list of benefits and potential harms of ginger use for the management of NVP. The authors suggest that this guideline should be addressed during the clinical consultation for NVP and should help to take a decision to use ginger or not. They reported that even if no conclusive evidence of adverse events of ginger on fetus was shown nowadays, the potential anti-coagulant effect of ginger is still equivocal and have to take account by clinicians before ginger recommendation.

Can we safely give ginger rhizome to decrease nausea and vomiting in women during early pregnancy?

First, medical supervision is mandatory. Ginger recommendation has to be done on a case-by-case basis after evaluation of patient medical history. Then, the quality of the finish product containing ginger and the quality of ginger itself, the quality of its transformation process (powder, extract, oils, etc.), and its relative standardization have to be mastered to assure consumer safety. If all the prerogatives are met, doctors could recommend ginger for NVP in early pregnancy.

At the least, further clinical studies are still needed to highlight the effect of ginger on platelet aggregation, particularly in pregnant women population.

**Acknowledgments:** The author(s) received no specific funding for this work.

**Author Contributions:** J.S., P.-Y.M. and S.L. made literature searches. S.L. analyzed the data and all authors discussed them. J.S. & S.L. wrote the manuscript, P.-Y.M. reviewed it.

**Conflicts of Interest:** Gynov SAS is involved in research/development and marketing/sales of Gynosea® as a food supplement for nausea and vomiting in early pregnancy. The general goal of Gynov SAS is to develop food supplements or FSMP with scientifically-proven well-being benefits. Gynosea® contains ginger rhizome extract, magnesium, and vitamin B6. Therefore, Gynov SAS has a commercial interest in this publication.

## Abbreviations

| | |
|---|---|
| T&CM | traditional & complementary medicine |
| CAM | complementary and alternative medicine |
| USDA | United States Department of Agriculture |
| INCA | Étude individuelle nationale des consommations alimentaires |
| ANSES | Agence nationale de sécurité sanitaire de l'alimentation, de l'environnement et du travail (France) |
| UK | United Kingdom |
| EU | European Union |
| USA | United States of America |
| RCT | randomized clinical trial |
| EC | European Commission |
| EFSA | European Food Safety Authority |
| NVP | nausea and vomiting of pregnancy |
| BW | Body Weight |
| ca. | circa |
| HMPC | Committee on Herbal Medicinal Products |

## References

1. Organisation mondiale de la santé (OMS). *Stratégie de L'oms Pour la Médecine Traditionnelle Pour 2014–2023*; Organisation mondiale de la santé: Genève, Suisse, 2013.
2. Parlement Européen et Conseil de l'Europe. Directive 2002/46/ce du Parlement Européen et du Conseil du 10 Juin 2002 Relative au Rapprochement des Législations des Etats Membres Concernant les Compléments Alimentaires. 2002. Available online: https://www.legifrance.gouv.fr/affichTexte.do?cidTexte=JORFTEXT000000337464&dateTexte=20090923 (accessed on 27 November 2017).
3. European Food Safety Authority (EFSA). Scientific report of EFSA—Compendium of botanicals reported to contain naturally occuring substances of possible concern for human health when used in food and food supplements. *EFSA J.* **2012**, *10*, 2663.
4. Ministère Croate [MINISTARSTVO ZDRAVLJA]. Liste positive vitamines, minéraux, plantes, substances à but physiologiques [pravilnik o tvarima koje se mogu dodavati hrani i koristiti u proizvodnji hrane te tvarima čije je korištenje u hrani zabranjeno ili ograničeno]. *Off. J.* **2013**, *160*, 13.
5. Etat Français. Arrêté du 24 Juin 2014 Etablissant la Liste des Plantes, Autres que les Champignons, Autorisées dans les Compléments Alimentaires et les Conditions de Leur Emploi. 2014. Available online: https://www.legifrance.gouv.fr/affichTexte.do?cidTexte=JORFTEXT000029254516&categorieLien=id (accessed on 27 November 2017).
6. Etat Belge; Etat Français; Etat Italien. Liste Belfrit—Harmonisation de L'emploi des Plantes dans les Compléments Alimentaires au sein D'un Espace Européen: Belgique, France, Italie. 2014. Available online: https://www.economie.gouv.fr/dgccrf/projet-belfrit-cooperation-reussie-au-sein-lunion-europeenne (accessed on 27 November 2017).
7. Etat Hongrois. Az oéti Szakértői Testülete Altal Etrend-Kiegészítőkben Alkalmazásra nem Javasolt Növények [Plantes Pour une Utilisation Dans les Compléments Alimentaires ne sont pas Recommandés par L'oéti Body Expert]. 2013. Available online: https://anzdoc.com/az-oeti-szakerti-testlete-altal-etrend-kiegeszitkben-alkalma.html (accessed on 27 November 2017).
8. Eardley, S.; Bishop, F.L.; Prescott, P.; Cardini, F.; Brinkhaus, B.; Santos-Rey, K.; Vas, J.; von Ammon, K.; Hegyi, G.; Dragan, S.; et al. A systematic literature review of complementary and alternative medicine prevalence in eu. *Forsch. Komplementmed.* **2012**, *19* (Suppl. 2), 18–28. [CrossRef] [PubMed]
9. Agence Nationale de Sécurité Sanitaire Alimentation, de L'alimentation, de L'environnement et du Travail (ANSES). Étude Individuelle Nationale des Consommations Alimentaires 3 (inca 3). 2017. Available online: https://www.anses.fr/fr/content/etude-inca-3-pr%C3%A9sentation (accessed on 29 November 2017).
10. De Ridder, K.; Bel, S.; Brocatus, L.; Lebacq, T.; Ost, C.; Teppers, E. *Enquête de consommation alimentaire 2014-2015*; Numéro de dépôt: D/2016/2505/51 Référence Interne: PHS Report 2016-042; Institut Scientifique de Santé Publique: Bruxelles, Belgique, 2016.
11. O'Brien, O.A.; Lindsay, K.L.; McCarthy, M.; McGloin, A.F.; Kennelly, M.; Scully, H.A.; McAuliffe, F.M. Influences on the food choices and physical activity behaviours of overweight and obese pregnant women: A qualitative study. *Midwifery* **2017**, *47*, 28–35. [CrossRef] [PubMed]
12. Skeie, G.; Braaten, T.; Hjartaker, A.; Lentjes, M.; Amiano, P.; Jakszyn, P.; Pala, V.; Palanca, A.; Niekerk, E.M.; Verhagen, H.; et al. Use of dietary supplements in the european prospective investigation into cancer and nutrition calibration study. *Eur J. Clin. Nutr.* **2009**, *63* (Suppl. 4), S226–S238. [CrossRef] [PubMed]
13. Birdee, G.S.; Kemper, K.J.; Rothman, R.; Gardiner, P. Use of complementary and alternative medicine during pregnancy and the postpartum period: An analysis of the national health interview survey. *J. Womens Health (Larchmt)* **2014**, *23*, 824–829. [CrossRef] [PubMed]
14. Holst, L.; Wright, D.; Haavik, S.; Nordeng, H. Safety and efficacy of herbal remedies in obstetrics–review and clinical implications. *Midwifery* **2011**, *27*, 80–86. [CrossRef] [PubMed]
15. Frawley, J.; Adams, J.; Sibbritt, D.; Steel, A.; Broom, A.; Gallois, C. Prevalence and determinants of complementary and alternative medicine use during pregnancy: Results from a nationally representative sample of australian pregnant women. *Aust. N. Z. J. Obstet. Gynaecol.* **2013**, *53*, 347–352. [CrossRef] [PubMed]
16. Hastings-Tolsma, M.; Vincent, D. Decision-making for use of complementary and alternative therapies by pregnant women and nurse midwives during pregnancy: An exploratory qualitative study. *Int. J. Nurs. Midwifery* **2013**, *5*, 76–89. [CrossRef]

17. Olowokere, A.; Olajide, O. Women's perception of safety and utilization of herbal remedies during pregnancy in a local government area in nigeria. *Clin. Nurs. Stud.* **2013**, *1*. [CrossRef]
18. Kennedy, D.A.; Lupattelli, A.; Koren, G.; Nordeng, H. Safety classification of herbal medicines used in pregnancy in a multinational study. *BMC Complement. Altern. Med.* **2016**, *16*, 102. [CrossRef] [PubMed]
19. Pallivalappila, A.R.; Stewart, D.; Shetty, A.; Pande, B.; Singh, R.; McLay, J.S. Complementary and alternative medicine use during early pregnancy. *Eur. J. Obstet. Gynecol. Reprod. Biol.* **2014**, *181*, 251–255. [CrossRef] [PubMed]
20. Stewart, D.; Pallivalappila, A.R.; Shetty, A.; Pande, B.; McLay, J.S. Healthcare professional views and experiences of complementary and alternative therapies in obstetric practice in north east scotland: A prospective questionnaire survey. *BJOG* **2014**, *121*, 1015–1019. [CrossRef] [PubMed]
21. Ahmed, M.; Hwang, J.H.; Choi, S.; Han, D. Safety classification of herbal medicines used among pregnant women in asian countries: A systematic review. *BMC Complement. Altern. Med.* **2017**, *17*, 489. [CrossRef] [PubMed]
22. CBI Market Intelligence; The Ministry of Foreign Affairs. CBI Product Factsheet: Dried Ginger in Europe. Available online: www.cbi.eu/market-information (accessed on 3 November 2017).
23. World Health Organization (WHO). *Who Monographs on Selected Medicinal Plants Volume 1*; World Health Organization (WHO): Geneva, Switzerland, 1999.
24. Bradley, P. British Herbal Compendium. In *A Handbook of Scientific Information on Widely Used Plant Drugs*; British Herbal Medicine Association: Bournemouth, UK, 1992; Volume 1.
25. European Scientific Cooperative on Phytotherapy (ESCOP). *Escop Monographs: The Scientific Foundation for Herbal Medicinal Products*, 2nd ed.; European Scientific Cooperative on Phytotherapy: Stuttgart, Germany, 2009; ISBN 9783131294210.
26. European Medicines Agency (EMA); Committee on Herbal Medicinal Products (HMPC). *Community Herbal Monograph on Zingiber Officinale Roscoe, Rhizoma*; European Medicines Agency: London, UK, 2012.
27. Khan, S.; Pandotra, P.; Qazi, A.; Lone, S.; Muzafar, M.; Gupta, A.; Gupta, S. Medicinal and nutritional qualities of zingiber officinale. In *Fruits, Vegetables, and Herbs*; Watson, R.R., Preedy, V.R., Eds.; Academic Press: Oxford, UK, 2016; pp. 525–550.
28. Kemper, K. Ginger (*zingiber officinale*); Longwood Herbal Task Force and the Center for Holistic Pediatric Education and Research. Longwood Herbal Task Force. 1999. Available online: http://www.mcp.edu/herbal/default.htm (accessed on 15 November 2017).
29. Dhanik, J.; Arya, N.; Nand, V. A review on zingiber officinale. *J. Pharmacogn. Phytochem.* **2017**, *6*, 174–184.
30. European Medicines Agency (EMA); Committee on Herbal Medicinal Products (HMPC). *Assessment Report on Zingiber Officinale Roscoe, Rhizome*; European Medicines Agency: London, UK, 2012.
31. Agence Nationale de Sécurité Sanitaire Alimentation, de L'alimentation, de L'environnement et du Travail (ANSES). Ciqual, Table de Composition Nutritionnelle des Aliments. Available online: https://ciqual.anses.fr/#/aliments/11006/gingembre-poudre (accessed on 6 November 2017).
32. Jolad, S.D.; Lantz, R.C.; Solyom, A.M.; Chen, G.J.; Bates, R.B.; Timmermann, B.N. Fresh organically grown ginger (*zingiber officinale*): Composition and effects on lps-induced pge2 production. *Phytochemistry* **2004**, *65*, 1937–1954. [CrossRef] [PubMed]
33. Jolad, S.D.; Lantz, R.C.; Chen, G.J.; Bates, R.B.; Timmermann, B.N. Commercially processed dry ginger (*zingiber officinale*): Composition and effects on lps-stimulated pge2 production. *Phytochemistry* **2005**, *66*, 1614–1635. [CrossRef] [PubMed]
34. Ali, B.H.; Blunden, G.; Tanira, M.O.; Nemmar, A. Some phytochemical, pharmacological and toxicological properties of ginger (*zingiber officinale* roscoe): A review of recent research. *Food Chem. Toxicol.* **2008**, *46*, 409–420. [CrossRef] [PubMed]
35. Sharifi-Rad, M.; Varoni, E.M.; Salehi, B.; Sharifi-Rad, J.; Matthews, K.R.; Ayatollahi, S.A.; Kobarfard, F.; Ibrahim, S.A.; Mnayer, D.; Zakaria, Z.A.; et al. Plants of the genus zingiber as a source of bioactive phytochemicals: From tradition to pharmacy. *Molecules* **2017**, *22*. [CrossRef] [PubMed]
36. Agrahari, P.; Panda, P.; Verma, N.; Khan, W.; Darbari, S. A brief study on zingiber officinale-a review. *J. Drug Discov. Ther.* **2015**, *3*, 20–27.
37. Petersen, I.; McCrea, R.L.; Lupattelli, A.; Nordeng, H. Women's perception of risks of adverse fetal pregnancy outcomes: A large-scale multinational survey. *BMJ Open* **2015**, *5*, e007390. [CrossRef] [PubMed]

38. Plengsuriyakarn, T.; Viyanant, V.; Eursitthichai, V.; Tesana, S.; Chaijaroenkul, W.; Itharat, A.; Na-Bangchang, K. Cytotoxicity, toxicity, and anticancer activity of zingiber officinale roscoe against cholangiocarcinoma. *Asian J. Cancer Prev.* **2012**, *13*, 4597–4606. [CrossRef]

39. Harliansyah; Murad, N.; Ngah, W.; Yusof, A. Antiproliferative, antioxidant and apoptosis effects of zingiber officinale and 6-gingerol on hepg2 cells. *Asian J. Biochem.* **2007**, *6*, 421–426.

40. Unnikrishnan, M.C.; Kuttan, R. Cytotoxicity of extracts of spices to cultured cells. *Nutr. Cancer* **1988**, *11*, 251–257. [CrossRef] [PubMed]

41. Abudayyak, M.; Ozdemir Nath, E.; Ozhan, G. Toxic potentials of ten herbs commonly used for aphrodisiac effect in turkey. *Turk. J. Med. Sci.* **2015**, *45*, 496–506. [CrossRef] [PubMed]

42. Zaeoung, S.; Plubrukarn, A.; Keawpradub, N. Cytotoxic and free radical scavenging activities of zingiberaceous rhizomes. *Songklanakarin J. Sci. Technol.* **2005**, *27*, 799–812.

43. Wei, Q.; Ma, J.; Cai, Y.; Yang, L.; Liu, Z. Cytotoxic and apoptotic activities of diarylheptanoids and gingerol-related compounds from the rhizome of chinese ginger. *J. Ethnopharmacol.* **2005**, *102*, 177–184. [CrossRef] [PubMed]

44. Yang, G.; Zhong, L.; Jiang, L.; Geng, C.; Cao, J.; Sun, X.; Ma, Y. Genotoxic effect of 6-gingerol on human hepatoma g2 cells. *Chem. Biol. Interact.* **2010**, *185*, 12–17. [CrossRef] [PubMed]

45. Kim, J.; Lee, S.; Park, H.; Yang, J.; Shin, T.; Kim, Y.; Baek, N.; Kim, S.; Choi, S.; Kwon, B.; et al. Cytotoxic components from the dried rhizomes of zingiber officinale roscoe. *Arch. Pharm. Res.* **2008**, *31*, 415–418. [CrossRef] [PubMed]

46. Peng, F.; Tao, Q.; Wu, X.; Dou, H.; Spencer, S.; Mang, C.; Xu, L.; Sun, L.; Zhao, Y.; Li, H.; et al. Cytotoxic, cytoprotective and antioxidant effects of isolated phenolic compounds from fresh ginger. *Fitoterapia* **2012**, *83*, 568–585. [CrossRef] [PubMed]

47. Zick, S.M.; Djuric, Z.; Ruffin, M.T.; Litzinger, A.J.; Normolle, D.P.; Alrawi, S.; Feng, M.R.; Brenner, D.E. Pharmacokinetics of 6-gingerol, 8-gingerol, 10-gingerol, and 6-shogaol and conjugate metabolites in healthy human subjects. *Cancer Epidemiol. Biomark. Prev.* **2008**, *17*, 1930–1936. [CrossRef] [PubMed]

48. Yu, Y.; Zick, S.; Li, X.; Zou, P.; Wright, B.; Sun, D. Examination of the pharmacokinetics of active ingredients of ginger in humans. *AAPS J.* **2011**, *13*, 417–426. [CrossRef] [PubMed]

49. Soudamini, K.K.; Unnikrishnan, M.C.; Sukumaran, K.; Kuttan, R. Mutagenicity and anti-mutagenicity of selected spices. *Indian J. Physiol. Pharmacol.* **1995**, *39*, 347–353. [PubMed]

50. Nagabhushan, M.; Amonkar, A.J.; Bhide, S.V. Mutagenicity of gingerol and shogaol and antimutagenicity of zingerone in salmonella/microsome assay. *Cancer Lett.* **1987**, *36*, 221–233. [CrossRef]

51. NIrmala, K.; Prasanna Krishna, T.; Polasa, K. In vivo antimutagenic potential of ginger on formation and excretion of urinary mutagens in rats. *Int. J. Cancer Res.* **2007**, *3*, 134–142.

52. Rong, X.; Peng, G.; Suzuki, T.; Yang, Q.; Yamahara, J.; Li, Y. A 35-day gavage safety assessment of ginger in rats. *Regul. Toxicol. Pharmacol.* **2009**, *54*, 118–123. [CrossRef] [PubMed]

53. Malik, Z.; Sharmaa, P. Attenuation of high-fat diet induced body weight gain, adiposity and biochemical anomalies after chronic administration of ginger (*zingiber officinale*) in wistar rats. *Int. J. Pharmacol.* **2011**, *7*, 801–812. [CrossRef]

54. Anonymous. Zingiberis rhizome. In *Escop Monographs*; Stuttgart, T.P., Ed.; European Scientific Cooperative on Phytotherapy: New York, NY, USA, 2003; pp. 547–553.

55. Weidner, M.; Sigwart, K. Investigation of the teratogenic potential of a zingiber officinale extract in the rat. *Reprod. Toxicol.* **2001**, *15*, 75–80. [CrossRef]

56. Tanabe, M.; Chen, Y.D.; Saito, K.; Kano, Y. Cholesterol biosynthesis inhibitory component from zingiber officinale roscoe. *Chem. Pharm. Bull. (Tokyo)* **1993**, *41*, 710–713. [CrossRef] [PubMed]

57. Jeena, K.; Liju, V.B.; Kuttan, R. A preliminary 13-week oral toxicity study of ginger oil in male and female wistar rats. *Int. J. Toxicol.* **2011**, *30*, 662–670. [CrossRef] [PubMed]

58. Ojewole, J.A. Analgesic, antiinflammatory and hypoglycaemic effects of ethanol extract of zingiber officinale (roscoe) rhizomes (zingiberaceae) in mice and rats. *Phytother. Res.* **2006**, *20*, 764–772. [CrossRef] [PubMed]

59. Suekawa, M.; Ishige, A.; Yuasa, K.; Sudo, K.; Aburada, M.; Hosoya, E. Pharmacological studies on ginger. I. Pharmacological actions of pungent constitutents, (6)-gingerol and (6)-shogaol. *J. Pharmacobiodyn.* **1984**, *7*, 836–848. [CrossRef] [PubMed]

60. Wu, H.; Ye, D.; Bai, Y.; Zhao, Y. Effect of dry ginger and roasted ginger on experimental gastric ulcers in rats. *China J. Chin. Mater. Medica* **1990**, *15*, 278–280, 317–318.

61. Wilkinson, J.M. Effect of ginger tea on the fetal development of sprague-dawley rats. *Reprod. Toxicol.* **2000**, *14*, 507–512. [CrossRef]

62. Sukandar, E.; Qowiyah, A.; Purnamasari, R. Teratogenicity study of combination of ginger rhizome extract and noni fruit extract in wistar rat. *Indones. J. Pharm.* **2009**, *20*, 48–54.

63. Fischer-Rasmussen, W.; Kjaer, S.K.; Dahl, C.; Asping, U. Ginger treatment of hyperemesis gravidarum. *Eur. J. Obstet. Gynecol. Reprod. Biol.* **1991**, *38*, 19–24. [CrossRef]

64. Chrubasik, S.; Pittler, M.H.; Roufogalis, B.D. Zingiberis rhizoma: A comprehensive review on the ginger effect and efficacy profiles. *Phytomedicine* **2005**, *12*, 684–701. [CrossRef] [PubMed]

65. Betz, O.; Kranke, P.; Geldner, G.; Wulf, H.; Eberhart, L.H. Is ginger a clinically relevant antiemetic? A systematic review of randomized controlled trials. *Res. Complement. Nat. Class. Med.* **2005**, *12*, 14–23. [CrossRef] [PubMed]

66. Haniadka, R.; Saldanha, E.; Sunita, V.; Palatty, P.L.; Fayad, R.; Baliga, M.S. A review of the gastroprotective effects of ginger (*zingiber officinale* roscoe). *Food Funct.* **2013**, *4*, 845–855. [CrossRef] [PubMed]

67. Viljoen, E.; Visser, J.; Koen, N.; Musekiwa, A. A systematic review and meta-analysis of the effect and safety of ginger in the treatment of pregnancy-associated nausea and vomiting. *Nutr. J.* **2014**, *13*, 20. [CrossRef] [PubMed]

68. Marx, W.; McKavanagh, D.; McCarthy, A.L.; Bird, R.; Ried, K.; Chan, A.; Isenring, L. The effect of ginger (*zingiber officinale*) on platelet aggregation: A systematic literature review. *PLoS ONE* **2015**, *10*, e0141119.

69. Wang, Y.; Yu, H.; Zhang, X.; Feng, Q.; Guo, X.; Li, S.; Li, R.; Chu, D.; Ma, Y. Evaluation of daily ginger consumption for the prevention of chronic diseases in adults: A cross-sectional study. *Nutrition* **2017**, *36*, 79–84. [CrossRef] [PubMed]

70. Sharifzadeh, F.; Kashanian, M.; Koohpayehzadeh, J.; Rezaian, F.; Sheikhansari, N.; Eshraghi, N. A comparison between the effects of ginger, pyridoxine (vitamin b6) and placebo for the treatment of the first trimester nausea and vomiting of pregnancy (nvp). *J. Matern. Fetal Neonatal Med.* **2017**, 1–6. [CrossRef] [PubMed]

71. Portnoi, G.; Chng, L.A.; Karimi-Tabesh, L.; Koren, G.; Tan, M.P.; Einarson, A. Prospective comparative study of the safety and effectiveness of ginger for the treatment of nausea and vomiting in pregnancy. *Am. J. Obstet. Gynecol.* **2003**, *189*, 1374–1377. [CrossRef]

72. Heitmann, K.; Nordeng, H.; Holst, L. Safety of ginger use in pregnancy: Results from a large population-based cohort study. *Eur J. Clin. Pharmacol.* **2013**, *69*, 269–277. [CrossRef] [PubMed]

73. Choi, J.S.; Han, J.Y.; Ahn, H.K.; Lee, S.W.; Koong, M.K.; Velazquez-Armenta, E.Y.; Nava-Ocampo, A.A. Assessment of fetal and neonatal outcomes in the offspring of women who had been treated with dried ginger (*zingiberis rhizoma* siccus) for a variety of illnesses during pregnancy. *J. Obstet. Gynaecol.* **2015**, *35*, 125–130. [CrossRef] [PubMed]

74. Paritakul, P.; Ruangrongmorakot, K.; Laosooksathit, W.; Suksamarnwong, M.; Puapornpong, P. The effect of ginger on breast milk volume in the early postpartum period: A randomized, double-blind controlled trial. *Breastfeed. Med.* **2016**, *11*, 361–365. [CrossRef] [PubMed]

75. Pongrojpaw, D.; Somprasit, C.; Chanthasenanont, A. A randomized comparison of ginger and dimenhydrinate in the treatment of nausea and vomiting in pregnancy. *J. Med. Assoc. Thai* **2007**, *90*, 1703–1709. [PubMed]

76. Saberi, F.; Sadat, Z.; Abedzadeh-Kalahroudi, M.; Taebi, M. Acupressure and ginger to relieve nausea and vomiting in pregnancy: A randomized study. *Iran. Red Crescent Med. J.* **2013**, *15*, 854–861. [CrossRef] [PubMed]

77. Saberi, F.; Sadat, Z.; Abedzadeh-Kalahroudi, M.; Taebi, M. Effect of ginger on relieving nausea and vomiting in pregnancy: A randomized, placebo-controlled trial. *Nurs. Midwifery Stud.* **2014**, *3*, e11841. [CrossRef] [PubMed]

78. Keating, A.; Chez, R.A. Ginger syrup as an antiemetic in early pregnancy. *Altern. Ther. Health Med.* **2002**, *8*, 89–91. [PubMed]

79. Smith, C.; Crowther, C.; Willson, K.; Hotham, N.; McMillian, V. A randomized controlled trial of ginger to treat nausea and vomiting in pregnancy. *Obstet. Gynecol.* **2004**, *103*, 639–645. [CrossRef] [PubMed]

80. Vutyavanich, T.; Kraisarin, T.; Ruangsri, R. Ginger for nausea and vomiting in pregnancy: Randomized, double-masked, placebo-controlled trial. *Obstet. Gynecol.* **2001**, *97*, 577–582. [CrossRef] [PubMed]

81. Chittumma, P.; Kaewkiattikun, K.; Wiriyasiriwach, B. Comparison of the effectiveness of ginger and vitamin b6 for treatment of nausea and vomiting in early pregnancy: A randomized double-blind controlled trial. *J. Med. Assoc. Thai.* **2007**, *90*, 15–20. [PubMed]
82. Ozgoli, G.; Goli, M.; Simbar, M. Effects of ginger capsules on pregnancy, nausea, and vomiting. *J. Altern. Complement. Med.* **2009**, *15*, 243–246. [CrossRef] [PubMed]
83. Ensiyeh, J.; Sakineh, M.A. Comparing ginger and vitamin b6 for the treatment of nausea and vomiting in pregnancy: A randomised controlled trial. *Midwifery* **2009**, *25*, 649–653. [CrossRef] [PubMed]
84. Mohammadbeigi, R.; Shahgeibi, S.; Soufizadeh, N.; Rezaiie, M.; Farhadifar, F. Comparing the effects of ginger and metoclopramide on the treatment of pregnancy nausea. *Pak. J. Biol. Sci.* **2011**, *14*, 817–820. [PubMed]
85. Willetts, K.E.; Ekangaki, A.; Eden, J.A. Effect of a ginger extract on pregnancy-induced nausea: A randomised controlled trial. *Aust. N. Z. J. Obstet. Gynaecol.* **2003**, *43*, 139–144. [CrossRef] [PubMed]
86. Basirat, Z.; Moghadamnia, A.; Kashifard, M.; Sharifi-Ravazi, A. The effect of ginger biscuit on nausea and vomiting in early pregnancy. *Acta Medica Iran.* **2009**, *47*, 51–56.
87. Boltman-Binkowski, H. A systematic review: Are herbal and homeopathic remedies used during pregnancy safe? *Curationis* **2016**, *39*, 1514. [CrossRef] [PubMed]
88. McLay, J.S.; Izzati, N.; Pallivalapila, A.R.; Shetty, A.; Pande, B.; Rore, C.; Al Hail, M.; Stewart, D. Pregnancy, prescription medicines and the potential risk of herb-drug interactions: A cross-sectional survey. *BMC Complement. Altern. Med.* **2017**, *17*, 543. [CrossRef] [PubMed]
89. Lete, I.; Allue, J. The effectiveness of ginger in the prevention of nausea and vomiting during pregnancy and chemotherapy. *Integr. Med. Insights* **2016**, *11*, 11–17. [CrossRef] [PubMed]
90. Ding, M.; Leach, M.; Bradley, H. The effectiveness and safety of ginger for pregnancy-induced nausea and vomiting: A systematic review. *Women Birth* **2013**, *26*, e26–e30. [CrossRef] [PubMed]
91. Thomson, M.; Corbin, R.; Leung, L. Effects of ginger for nausea and vomiting in early pregnancy: A meta-analysis. *J. Am. Board Fam. Med.* **2014**, *27*, 115–122. [CrossRef] [PubMed]
92. Campbell, K.; Rowe, H.; Azzam, H.; Lane, C.A. The management of nausea and vomiting of pregnancy. *J. Obstet. Gynaecol. Can.* **2016**, *38*, 1127–1137. [CrossRef] [PubMed]
93. Schwertner, H.A.; Rios, D.C.; Pascoe, J.E. Variation in concentration and labeling of ginger root 39 dietary supplements. *Obstet. Gynecol.* **2006**, *107*, 1337–1343. [CrossRef] [PubMed]
94. Mukkavilli, R.; Yang, C.; Tanwar, R.S.; Saxena, R.; Gundala, S.R.; Zhang, Y.; Ghareeb, A.; Floyd, S.D.; Vangala, S.; Kuo, W.W.; et al. Pharmacokinetic-pharmacodynamic correlations in the development of ginger extract as an anticancer agent. *Sci. Rep.* **2018**, *8*, 3056. [CrossRef] [PubMed]
95. European Medicines Agency (EMA); Committee on herbal medicinal products (HMPC). *Opinion of the HPMC on a Community Herbal Monograph on Zingiber Officinale Roscoe, Rhizome*; EMA/HPMC/216956/2012; European Medicines Agency (EMA): London, UK, 2012.
96. Shawahna, R.; Taha, A. Which potential harms and benefits of using ginger in the management of nausea and vomiting of pregnancy should be addressed? A consensual study among pregnant women and gynecologists. *BMC Complement. Altern. Med.* **2017**, *17*, 204. [CrossRef] [PubMed]

*foods*

MDPI

*Article*

# Systematic Review and Meta-Analysis of a Proprietary Alpha-Amylase Inhibitor from White Bean (*Phaseolus vulgaris* L.) on Weight and Fat Loss in Humans

Jay Udani *, Ollie Tan and Jhanna Molina

Medical Private Practice, Agoura Hills, CA 91301, USA; ollietan124@yahoo.com (O.T.);
jhannamolina@gmail.com (J.M.)
* Correspondence: drjayu2016@gmail.com; Tel.: +1-818-885-8500

Received: 14 February 2018; Accepted: 12 April 2018; Published: 20 April 2018

**Abstract:** The aim of this meta-analysis was to examine the evidence for the effectiveness of a proprietary alpha-amylase inhibitor from white bean (*Phaseolus vulgaris* L.) supplementation interventions in humans on modification of body weight and fat mass. A systematic literature search was performed using three databases: PubMed, the Cochrane collaboration, and Google Scholar. In addition, the manufacturer was contacted for internal unpublished data, and finally, the reference section of relevant original research and review papers were mined for additional studies. Eleven studies were selected for the meta-analysis of weight loss (a total of 573 subjects), and three studies for the meta-analysis of body fat reduction (a total of 110 subjects), as they fulfilled the inclusion criteria. *Phaseolus vulgaris* supplementation showed an average effect on weight loss difference of $-1.08$ kg (95% CI (confidence interval), $-0.42$ kg to $-1.16$ kg, $p < 0.00001$), and the average effect on body fat reduction was 3.26 kg (95% CI, $-2.35$ kg to $-4.163$ kg, $p = 0.02$). This meta-analysis found statistically significant effects of *Phaseolus vulgaris* supplementation on body weight and body fat.

**Keywords:** weight loss; fat loss; obesity; alpha-amylase inhibitor; meta-analysis; *Phaseolus vulgaris* L.

## 1. Introduction

There are many dietary interventions available to counteract the epidemic of overweight and obesity which play a major role in the development of insulin resistance and type 2 diabetes mellitus [1]. One strategy is based on lowering the excessive intake of carbohydrates, especially, the refined ones [2]. This could be achieved by lowering the portions or replacing the carbohydrates with more fats or by adding soluble fiber to the diet which is thought to slow down the absorption of carbohydrates [3]. Lowering the glycemic index through the usage of fiber in the diet is not favored by most people due to potential taste preferences and adverse reactions resulting in gastrointestinal problems such as gas and diarrhea. Therefore, another strategy becomes more and more promising to impact the carbohydrate absorption by using bioactive ingredients which block or slow the carbohydrate absorption in the gastrointestinal tract via inhibiting the necessary enzymes, amylase and glucosidase [4]. Amylase breaks down complex carbohydrates, such as starch, into oligosaccharides and glucosidase enzymes further convert these to monosaccharides.

There are the different forms of amylase inhibitors, namely, Alpha-amylase inhibitor isoform 1 (Alpha-AI1), Alpha-AI2, and Alpha-AIL which can be found in in the embryonic axes and cotyledons in the seed of common beans (*Phaseolus* spp.) [5]. These so-called glycoproteins bind to alpha-amylase non-covalently, mainly through hydrophobic interaction, by completely blocking access to the active site of the alpha-amylase [6,7]. The Alpha-AI1 isoform is the one with anti-amylase bioactivity in humans, and therefore, inhibits the starch digestion [8]. This blocking affect is also dependent on pH, temperature, incubation time and the presence of particular ions which have been optimized for the specific and proprietary product named Phase2® brand *Phaseolus vulgaris* White Bean product (Pharmachem Laboratories, Kearny, NJ, USA) [9,10]. This particular dietary supplement has demonstrated its potential and ability to cause weight loss in numerous clinical trials in humans [11].

Phase2® brand *Phaseolus vulgaris* White Bean extract is made by a standardized water extract of non-GMO (Genetically Modified Organism) whole dried beans (*Phaseolus vulgaris*) which is made through a proprietary process. The white-to-beige powder consists of *Phaseolus vulgaris* (~90%) and Gum Arabic (~10%). It has at least 3000 alpha-amylase inhibiting units (AAIU) per gram when tested at a pH 6.8 using potato starch as the substrate and pancreatin as the enzyme source. The Phase2® brand products are used in dietary supplements in various forms, including powders, tablets, capsules and chewables for the application of weight control and weight loss. In addition, it is also incorporated in food products like chewing gum, mashed potatoes, yeast-raised dough (bread, pizza, etc.) without losing bioactivity or changing the appearance, texture or taste of the food [12–14].

The aim of this meta-analysis was to examine the evidence for the effectiveness of a proprietary alpha-amylase inhibitor from white bean (*Phaseolus vulgaris*) supplementation interventions on modification of body weight and fat mass.

## 2. Methods

This review was performed according to the PRISMA (preferred reporting items for systematic reviews and meta-analyses) statement for quality of reporting a meta-analysis [15].

### 2.1. Literature Search

Literature searches in PubMed, the Cochrane collaboration, and Google Scholar were undertaken using the following keywords: *Phaseolus vulgaris*, Alpha-amylase inhibitor/inhibition, Phase2®, White bean extract, kidney bean, starch blocker, weight loss, body weight, body fat, BMI (body mass index), anthropometric measures, obesity, overweight and safety.

In addition, the manufacturer was contacted for internal unpublished data, and finally, the reference section of relevant original research and review papers were mined for additional studies. No age, sex, geographic, time or publication status restrictions were imposed on the initial search.

### 2.2. Study Selection Criteria

Studies were eligible for inclusion if they met the following PICOS' (Participants, Intervention, Control, Outcome measurements, and Study design) criteria: (a) Participants: overweight or obese individuals; (b) Intervention: Phase2® brand *Phaseolus vulgaris* white bean extract, at least 1200 mg per day, for at least 4 weeks; (c) Control: studies comparing the experimental group (*Phaseolus vulgaris* supplementation) with a control/placebo group (no *Phaseolus vulgaris* supplementation ), or against baseline; (d) Outcome measurements: studies needed to include measurements of body mass or fat mass; (e) Study design: Studies needed to be either randomized, double-blind, placebo-controlled parallel or crossover trial, or open-label studies.

### 2.3. Assessment of Risk of Bias

For the quality assessment of randomized controlled trials (RCTs), we used the Delphi list, which includes eight questions with three response options "yes", "no", or "do not know" depending on compliance with key methodological components, and produces a quality score of maximum 9 points that provides an overall estimate of RCT quality [16].

## 2.4. Data Extraction and Quality Assessment

Two reviewers independently extracted the following data from the selected articles: publication year, number of participants (*Phaseolus vulgaris* and control group), baseline characteristics of the participants, methodological characteristics of the study, pre- and post-values and standard deviation for body mass and fat mass, and statistical information.

## 2.5. Statistical Analysis

A meta-analysis to estimate the overall treatment effect of *Phaseolus vulgaris* supplementation relative to control groups was performed. Standardized Mean Difference (d) was used in the determination of effect size. In getting the Standardized Mean Difference (d), Cramer's $v$ (which shows the magnitude size) and 95% CI (confidence interval) for each d was computed. The following guideline was used in reading magnitude effect size for Cramer's $v$: $v < 0.1$ small effect, $0.1 < v < 0.3$ medium effect, $v > 0.3$ large effect. Weighted effect size on all studies was done using the Hunter–Schmidt approach. Weights are objectively assigned based from sample sizes of the studies. Hence, studies with bigger sample size ($n = 60$ [17]) had higher weight, compared to studies with smaller sample size ($n = 10$ [18]). *p*-values of individual studies are transformed (logarithmic) and aggregated using Chi-square.

## 3. Results

### 3.1. Article Selection

One hundred and sixty-five articles were identified. From this list, 54 human studies were identified, and 5 of these were determined to be duplicates, leaving 49 unique studies. Of these 49 studies, only 13 involved the Phase2® *Phaseolus vulgaris* ingredient and one of the following outcomes: weight loss, body fat loss, or anthropometric measures (reductions in waist, hip or thigh measurements). A meta-analysis which included studies which did and did not use the Phase2® *Phaseolus vulgaris* ingredient was excluded. In addition, one of the studies was excluded as it did not provide means, standard deviations, nor *p*-values [19]. The meta-analysis for weight loss includes 11 studies (see prisma flow diagram, Figure 1), with a total of 573 subjects (see Table 1). The meta-analysis for fat loss includes 3 studies [18,20,21] with a total of 110 subjects (see Table 2). Three studies were excluded, as they did not measure fat mass [22–24]. One study was excluded as it only reported fat loss in percent and not in kilogram [25].

**Table 1.** Effects of *Phaseolus vulgaris* on body weight. The overall *p*-value was determined using Chi-square (Chi-square value (W) = 80.02).

| Study | Treatment Group | | | Control Group | | | *p* | Weight | Effect (d) | Weighted Mean Difference (Fixed) 95% CI | |
|---|---|---|---|---|---|---|---|---|---|---|---|
| | *n* | Mean | SD | *n* | Mean | SD | | | | Lower | Upper |
| Udani et al. 2007 [11] | 13 | −6.0 | | 12 | −4.7 | | 0.424 | 4% | −0.33 | −0.46 | 1.12 |
| Asano [26] | 9 | −2.9 | | | | | | 3% | −0.19 | −1.06 | 0.67 |
| Koike et al. 2005 [18] | 10 | −1.8 | | | | | 0.002 | 3% | −1.61 | −0.61 | −2.62 |
| Grube et al. 2014 [17] | 60 | −2.9 | 2.6 | 60 | −0.9 | 2.0 | 0.001 | 19% | −0.85 | −0.48 | −1.12 |
| Osorio et al. 2009 [23] | 49 | −2.3 | | 49 | 2.21 | | 0.001 | 15% | −1.00 | −0.41 | −1.60 |
| Rothacker (week 12) 2003 [27] | 30 | −6.9 | | 60 | 0.8 | | 0.029 | 9% | −0.58 | −0.62 | −1.09 |
| Wu et al. 2010 [24] | 50 | −1.9 | −0.2 | 51 | −0.4 | −0.1 | 0.049 | 15% | −0.40 | -0.00 | −0.79 |
| Celleno et al. 2007 [20] | 20 | −2.9 | −1.2 | 30 | −0.4 | 0.4 | | 9% | −2.99 | −2.25 | −3.73 |
| Thom et al. 2000 [21] | 20 | −3.5 | | 20 | 2.0 | | 0.001 | 6% | −1.13 | −0.46 | −0.12 |
| Udani et al. 2004 [28] | 20 | −3.8 | | 19 | −1.65 | | 0.35 | 6% | −0.30 | −0.33 | −0.93 |
| Yamada et al. [25] | 33 | −0.8 | 0.2 | 33 | | | 0.01 | 10% | −0.97 | −0.24 | −1.68 |
| Total | 314 | | | 259 | | | 0.001 | 100% | −1.08 | −0.43 | −1.16 |

SD: standard deviation, CI: confidence interval.

**Table 2.** Effects of *Phaseolus vulgaris* on body fat. The overall *p*-value was determined using Chi-square (Chi-square value (W) = 36.84).

| Study | Treatment Group | | | Control Group | | | *p* | Weight | Effect (d) | Weighted Mean Difference (Fixed) 95% CI | |
|---|---|---|---|---|---|---|---|---|---|---|---|
| | *n* | Mean | SD | *n* | Mean | SD | | | | Lower | Upper |
| Koike et al. 2005 [18] | 10 | −1.2 | −0.4 | | | | 0.001 | 17% | −1.58 | −2.58 | −0.57 |
| Celleno et al. 2007 [20] | 30 | −2.4 | −0.67 | 30 | −0.16 | −0.33 | 0.001 | 50% | −4.24 | −5.15 | −3.33 |
| Thom et al. 2000 [21] | 20 | −2.3 | −1.5 | 20 | 0.7 | −0.6 | 0.01 | 33% | −2.63 | −3.47 | −1.78 |
| Total | 60 | | | 50 | | | 0.02 | 100% | −3.26 | −4.16 | −2.35 |

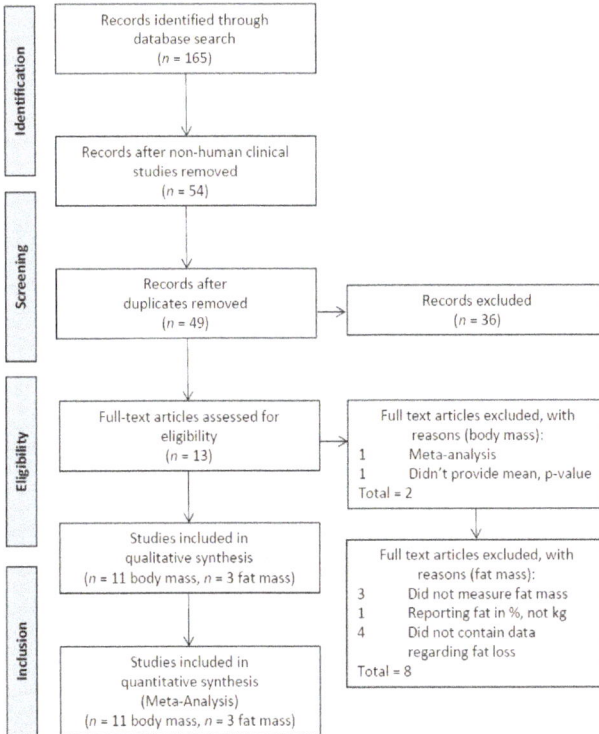

**Figure 1.** Prisma flow diagram for *Phaseolus vulgaris* and body mass, fat mass.

### 3.2. Phaseolus vulgaris Doses and Duration of Supplementation

The most common dose of *Phaseolus vulgaris* was 3000 mg per day, divided into three doses of 1000 mg (6/11). One study used 3000 mg per day, divided into two doses of 1500 mg [18], and one study used 2000 mg per day, divided into two doses of 1000 mg [11]. Two studies used 400 mg [21], and 445 mg [20] *Phaseolus vulgaris*, respectively, as part of a multi-ingredient blend. One study did not specify the amount of *Phaseolus vulgaris* used [25]. *Phaseolus vulgaris* was supplemented for 1 month (3/11) [11,20,23], 2 months (5/11) [18,24–26,28], or 3 months (3/11) (see Table 3) [17,21,27].

**Table 3.** Characteristics of the 11 clinical studies included in the meta-analysis.

| Study | Participants | | Intervention | | | n Phase 2 | Comparison | Methods | | Study Design | |
|---|---|---|---|---|---|---|---|---|---|---|---|
| | Country | Subjects Information | Dose | Diet Intervention | Duration of Intervention | | n Control | Weight | Fat Mass | Design | Delphi-Score |
| Asano et al. [26] | Japan | 5:1 female to male ratio; average age 36.3 ± 12.7; BMI > 25; average BMI = 31.6 | 3000 mg per day (1000 mg per meal) | no caloric restriction | 2 months | 9 | 0 | Scale | n/a | Open-Label | 2 |
| Udani et al. 2007 [22] | USA | 0.3:1 female to male ratio; age 18–40; average BMI = 26 | 2000 mg per day (1000 mg at breakfast & lunch) | maintain a caloric intake of 1800 per day | 4 weeks | 13 | 12 | Scale | - | RDBPC | 8 |
| Koike et al. 2005 [8] | Japan | 1:1 female to male ratio; mean age 41.1 and BMI range 23–30 | 2× per day 1500 mg Phase 2, 400 mg Clove, 40 mg Lysine 40 mg, 40 mg Arginine, 40 mg Alanine | no caloric restriction | 8 weeks | 10 | 0 | Scale | n/a | Open-Label | 1 |
| Grube et al. 2014 [17] | Germany | 3:1 female to male ratio; mean age 46; BMI range 25–35 | 3000 mg per day (1000 mg per meal) | hypocaloric (500 kcal), providing 40% of energy as carbohydrates | 12 weeks | 60 | 57 | Scale | BIA | RDBPC | 9 |
| Osorio et al. 2009 [23] | Mexico | obese and overweight (age range: 18–75 years) | 3000 mg per day (1000 mg per meal) | no caloric restriction besides carbohydrate-rich meals | 30 days | 37 | 0 | Scale | - | Open-Label | 1 |
| Rothacker 2003 [27] | USA | 24 male; 36 female; mean age 33.2; BMI range 24–32 | 3000 mg per day (1000 mg per meal) | no caloric restriction | 12 weeks | 34 | 26 | Scale | BIA | RDBPC | 8 |
| Wu et al. 2010 [24] | China | 1:1 female to male ratio; age 20–50; BMI range 25–40 | 3000 mg per day (1000 mg per meal) | no caloric restriction | 8 weeks | 51 | 50 | Scale | - | RDBPC | 8 |
| Celleno et al. 2007 [20] | Italy | 2.5:1 female to male ratio; mean age 34; average BMI = 26 | 3× per day 445 mg of Phase 2, 56 mg vitamin B3, and 0.5 mg chromium | carbohydrate-rich meals (100–200g) | 30 days | 30 | 29 | Scale | BIA | RDBPC | 8 |
| Thom et al. 2000 [21] | Norway | 9:1 female to male ratio; mean age 45.6; average BMI = 31 | 3× per day 400 mg Phase 2, 400 mg inulin, and 100 mg Garcinia cambogia | no caloric restriction | 12 weeks | 20 | 20 | Scale | BIA | RDBPC | 8 |
| Udani et al. 2004 [28] | USA | 9:1 female to male ratio; mean age 36.5; average weight of 193.1 pounds | 3000 mg per day (1000 mg per meal) | no caloric restriction | 8 weeks | 20 | 19 | Scale | BIA | RDBPC | 8 |
| Yamada et al. [25] | Japan | 1:1 female to male ratio; age 25–60; no BMI information | Twice a day proprietary functional food containing Phase 2 | no caloric restriction | 8 weeks | 23 | 24 | Scale | n/a | Open-Label | 4 |

BMI: body mass index; RDBPC: Randomized, Double-Blind, Placebo-Controlled; n/a: method not described; BIA: bioelectrical impedance analysis; -: not measured.

### 3.3. Control Groups

Participants of the control group had similar characteristics to the intervention groups, but they did not receive *Phaseolus vulgaris* supplementation. In most studies, placebo capsules were administered.

### 3.4. Effects on Body Weight

Table 1 summarizes the effects of *Phaseolus vulgaris* on body mass. *Phaseolus vulgaris* supplementation showed an average effect on weight loss difference of −1.08 kg (95% CI, −0.42 kg to −1.16 kg, $p < 0.00001$).

### 3.5. Effects on Fat Loss

Table 2 summarizes the effects of *Phaseolus vulgaris* on fat mass. The average effect of *Phaseolus vulgaris* supplementation on body fat reduction was 3.26 kg (95% CI, −2.35 kg to −4.163 kg, $p = 0.02$).

### 3.6. Risk of Bias and Publication Bias

Table 4 shows Delphi scores of each reviewed study. The Delphi scores varied between 1 and 9, the mean being 5.9 and the standard deviation ±3.2. Four studies obtained a score below the mean: Asano [26], Koike et al. 2005 [18], Osorio et al. 2009 [23], and Yamada et al. [25], with Asano [26], Koike et al. [18] and Osorio et al. [23] being open-label studies.

Table 4. Delphi-Scores.

| Delphi-Scores/Studies | Asano et al. [26] | Udani et al. 2007 [11] | Grube et al. 2014 [17] | Rothacker et al. 2003 [27] | Wu et al. 2010 [24] | Celleno et al. 2007 [20] | Thom et al. 2000 [21] | Udani et al. 2004 [28] | Koike et al. 2005 [18] | Osorio et al. 2009 [23] | Yamada et al. [25] |
|---|---|---|---|---|---|---|---|---|---|---|---|
| 1. Treatment allocation (a) Was a method of randomization performed? | 0 | 1 | 1 | 1 | 1 | 1 | 1 | 1 | 0 | 0 | 1 |
| (b) Was the treatment allocation concealed? | 0 | 1 | 1 | 1 | 1 | 1 | 1 | 1 | 0 | 0 | 0 |
| 2. Were the groups similar at baseline regarding the most important prognostic indicators? | 0 | 1 | 1 | 1 | 1 | 1 | 1 | 1 | 0 | 0 | 1 |
| 3. Where the eligibility criteria specified? | 1 | 1 | 1 | 1 | 1 | 1 | 1 | 1 | 0 | 0 | 1 |
| 4. Was the outcome assessor blinded? | 0 | 1 | 1 | 1 | 1 | 1 | 1 | 1 | 0 | 0 | 0 |
| 5. Was the care provider blinded? | 0 | 1 | 1 | 1 | 1 | 1 | 1 | 1 | 0 | 0 | 0 |
| 6. Was the patient blinded? | 0 | 1 | 1 | 1 | 1 | 1 | 1 | 1 | 0 | 0 | 0 |
| 7. Were point estimates and measures of variability presented for the primary outcome measures? | 1 | 1 | 1 | 1 | 1 | 1 | 1 | 1 | 1 | 1 | 1 |
| 8. Did the analysis include an intention-to-treat analysis? | 0 | 0 | 1 | 0 | 0 | 0 | 0 | 0 | 0 | 0 | 0 |
| Total Delphi Score | 2 | 8 | 9 | 8 | 8 | 8 | 8 | 8 | 1 | 1 | 4 |

## 4. Discussion

The aim of this meta-analysis was to determine the effectiveness of Phase2® (*Phaseolus vulgaris*) to support weight loss and to reduce body fat. The overall meta-analysis revealed a significant difference in change in body weight, and body fat between Phase2® (*Phaseolus vulgaris*) and placebo.

Barret et al. conducted a review of clinical studies with Phase 2 brand *Phaseolus vulgaris* White Bean product on weight loss and glycemic control [29]. The analysis identified ten clinical studies which have demonstrated weight loss over time following administration of Phase 2 when taken concurrently with meals containing carbohydrates. Three of these clinical studies revealed significant loss of body weight with Phase 2 compared to a placebo control in people who are overweight or obese. In addition, three clinical trials showed a reduction in serum triglycerides over time. Nine of these clinical studies reported by Barret et al. have been used in this systematic review and meta-analysis. The study by Vinson et al. was not taken into consideration due to its focus on glycemic index and blood glucose investigations without looking into weight loss parameters [30].

While our meta-analysis revealed a significant difference in weight loss over placebo, a previous meta-analysis of *Phaseolus vulgaris* [31] showed a non-significant difference in weight loss between *Phaseolus vulgaris* and placebo groups. This can be explained by the fact that the Onakpoya et al. meta-analysis included not only studies performed with Phase 2, but all studies on *Phaseolus vulgaris*. Both meta-analyses showed significant effects on fat loss. The importance of this work was to isolate the effects of the Phase 2 brand *Phaseolus vulgaris* White Bean from the body of literature. By using unpublished data and all arms of all studies available to us, we were able to demonstrate statistically significant effects on weight and body fat. Part of this importance is in the supplement industry, there is an assumed "generic equivalence" which we know to be false. The prior meta-analysis assumed such a generic equivalence and therefore came up with negative results. In this case, by limiting to only Phase 2, it does appear that there is significant weight and body fat loss with an excellent safety profile.

Low carbohydrate diets have been linked to weight loss, even when not consciously restricting calories, improved triglyceride levels, a reduction in blood glucose levels and improved insulin sensitivity, a decrease in blood pressure. Very low carbohydrate diets (ketogenic diets), with fewer than 50 g of carbohydrate per day, have been linked to weight loss and specific health benefits including neurological disorders. Adaptations to a ketogenic diet is often difficult and nutritional aids have been shown to be useful for entering into nutritional ketosis [32]. A recently concluded study indicated that both low-fat and low-carb diets can work for weight loss, and that there is no "best diet" when it comes to low-carb vs. low-fat diets. In total, 263 males and 346 premenopausal females were assigned to either a low-fat diet or a low-carb diet for 12 months. At 12 months, the low-fat group had lost 5.3 kg and the low-carb group 6.0 kg and this difference is neither statistically significant nor clinically relevant. The healthy diet that will work for you is the one you can stick to, and that varies by individual [33].

## 5. Conclusions

This meta-analysis found significant effect of Phase2® *Phaseolus vulgaris* supplementation on body weight and body fat.

**Acknowledgments:** This review was financially supported by Pharmachem Laboratories, Kearny, NJ, USA.

**Author Contributions:** J.U. was the primary author and was responsible for the design of the review and meta-analysis. J.M. was responsible for organizing references and extracting data from references into evidence tables. O.T. was responsible for the design and execution of meta-analyses. All authors reviewed and approved the final manuscript.

**Conflicts of Interest:** J.U. has received consulting fees from Pharmachem Laboratories for clinical trial and consulting work in the past. No fees were received for the writing or publication of this manuscript. The other authors have no conflicts of interest.

## References

1. De Toro-Martín, J.; Arsenault, B.J.; Després, J.P.; Vohl, M.C. Precision Nutrition: A Review of Personalized Nutritional Approaches for the Prevention and Management of Metabolic Syndrome. *Nutrients* **2017**, *9*, 913. [CrossRef] [PubMed]
2. Preuss, H.G.; Bagchi, D. Nutritional therapy of impaired glucose tolerance and diabetes mellitus. In *Nutritional Aspects and Clinical Management of Chronic Disorders and Diseases*; Bronner, F., Ed.; CRC Press: Boca Raton, FL, USA, 2002; pp. 69–91. ISBN 9781420041286.
3. Bell, S.J.; Sears, B. Low-Glycemic-Load Diets: Impact on Obesity and Chronic Diseases. *Crit. Rev. Food Sci. Nutr.* **2003**, *43*, 357–377. [CrossRef] [PubMed]
4. Fukagawa, N.K.; Anderson, J.W.; Hageman, G.; Young, V.R.; Ninaker, K.L. High-carbohydrate, high-fiber diets increase peripheral insulin sensitivity in healthy young and old adults. *Am. J. Clin. Nutr.* **1990**, *52*, 524–528. [CrossRef] [PubMed]
5. Bowman, D.E. Amylase inhibitor of navy bean. *Science* **1945**, *102*, 358–359. [CrossRef] [PubMed]
6. Santimone, M.; Koukiekolo, R.; Moreau, Y.; Le Berre, V.; Rouge, P.; Marchis-Mouren, G.; Desseaux, V. Porcine pancreatic a-amylase inhibition by the kidney bean (*Phaseolus vulgaris*) inhibitor (α-AI1) and structural changes in the α-amylase inhibitor complex. *Biochim. Biophys. Acta* **2004**, *1696*, 181–190. [CrossRef] [PubMed]
7. Bompard-Gilles, C.; Rousseau, P.; Rouge, P.; Payan, F. Substrate mimicry in the active center of a mammalian α-amylase: Structural analysis of an enzyme-inhibitor complex. *Structure* **1996**, *4*, 1441–1452. [CrossRef]
8. Payan, F. Structural basis for the inhibition of mammalian and insect α-amylases by plant protein inhibitors. *Biochim. Biophys. Acta* **2004**, *1696*, 171–180. [CrossRef] [PubMed]
9. Le Berre-Anton, V.; Bompard-Gilles, C.; Payan, F.; Rouge, P. Characterization and functional properties of the α-amylase inhibitor (α-AI) from kidney bean (*Phaseolus vulgaris*) seeds. *Biochim. Biophys. Acta* **1997**, *1343*, 31–40. [CrossRef]
10. Lajolo, F.M.; Finardi-Filho, F. Partial characterization of the amylase inhibitor of black beans (*Phaseolus vulgaris*), variety Rico 23. *J. Agric. Food Chem.* **1985**, *33*, 132–138. [CrossRef]
11. Udani, J.; Hardy, M.; Kavoussi, B. Dietary supplement carbohydrate digestion inhibitors: A review of the literature. In *Obesity. Epidmiology, Pathophysiology, and Prevention*; Bagchi, D., Preuss, H.G., Eds.; CRC Press: Boca Raton, FL, USA, 2007; pp. 279–298.
12. Udani, K. *The Mighty Bean*; European Baker: Bucharest, Romania, 2005.
13. Phase 2®/StarchLite. Available online: www.phase2info.com/pdf/Phase2_Study13.pdf (accessed on 5 January 2018).
14. Phase 2®/StarchLite in Chewing Gum. Available online: http://www.phase2info.com/pdf/Phase2_Study14.pdf (accessed on 5 January 2018).
15. Moher, D.; Liberati, A.; Tetzlaff, J.; Altman, D.G.; the PRISMA Group. Preferred Reporting Items for Systematic Reviews and Meta-Analyses: The PRISMA Statement. *Ann. Intern. Med.* **2009**, *151*, 264–269. [CrossRef] [PubMed]
16. Verhagen, A.P.; de Vet, H.C.W.; de Bie, R.A.; Kessels, A.G.; Boers, M.; Bouter, L.M.; Knipschild, P.G. The Delphi list: A criteria list for quality assessment of randomized clinical trials for conducting systematic reviews developed by Delphi consensus. *J. Clin. Epidemiol.* **1998**, *51*, 1235–1241. [CrossRef]
17. Grube, B.; Chong, W.; Chong, P.; Riede, L. Weight Reduction and Maintenance with IQP-PV-101: A 12-Week Randomized Controlled Study with a 24-Week Open Label Period. *Obesity* **2014**, *22*, 645–651. [CrossRef] [PubMed]
18. Koike, T.; Koizumi, Y.; Tang, L.; Takahara, K.; Saitou, Y. The antiobesity effect and the safety of taking "Phaseolamin(TM) 1600 diet". *J. New Rem. Clin.* **2005**, *54*, 1–16.
19. Erner, S.; Meiss, D. The Effect of Thera-Slim on Weight, Body Composition and Select Laboratory Parameters in Adults with Overweight and Mild-Moderate Obesity. Unpublished work. 2003.
20. Celleno, L.; Tolaini, M.V.; D'Amore, A.; Perricone, N.V.; Preuss, H.G. A dietary supplement containing standardized Phaseolus vulgaris extract influences body composition of overweight men and women. *Int. J. Med. Sci.* **2007**, *49*, 45–52. [CrossRef]
21. Thom, E. A randomized, double-blind, placebo-controlled trial of a new weight-reducing agent of natural origin. *J. Int. Med. Res.* **2000**, *28*, 229–233. [CrossRef] [PubMed]

22. Udani, J.; Singh, B. Blocking Carbohydrate Absorption and Weight Loss: A Clinical Trial Using a Proprietary Fractionated White Bean Extract. *Altern. Ther. Health Med.* **2007**, *13*, 32–37. [PubMed]
23. Osorio, L.; Gamboa, J. Random Multi-Center Evaluation to Test the Efficacy of *Phaseolus vulgaris* (Precarb) in Obese and Overweight Individuals. Unpublished work. 2005.
24. Wu, X.; Xiaofeng, X.; Shen, J.; Perricone, N.; Preuss, H. Enhanced Weight Loss From a Dietary Supplement Containing Standardized *Phaseolus vulgaris* Extract in Overweight Men and Women. *J. Appl. Res.* **2010**, *10*, 73–79.
25. Yamada, J.; Yamamoto, T.; Hideyo, Y. Effects of Combination of Functional Food Materials on Body Weight, Body Fat Percentage, Serum Triglyceride and Blood Glucose. Unpublished work. 2007.
26. Asano, N. The Report of the Test Regarding the Efficacy of Phaseolamine on Weight Loss. Unpublished work, 2009.
27. Rothacker, D. Reduction in body weight with a starch blocking diet aid: StarchAway comparison with placebo. *Leiner Health Prod.* **2003**. Unpublished work.
28. Udani, J.; Hardy, M.; Madsen, D.C. Blocking carbohydrate absorption and weight loss: A clinical trial using Phase 2 brand proprietary fractionated white bean extract. *Altern. Med. Rev.* **2004**, *99*, 63–69.
29. Barrett, M.L.; Udani, J.K. A proprietary alpha-amylase inhibitor from white bean (*Phaseolus vulgaris*): A review of clinical studies on weight loss and glycemic control. *Nutr. J.* **2011**, *10*, 24. [CrossRef] [PubMed]
30. Vinson, J.; Al Kharrat, H.; Shuta, D. Investigation of an amylase inhibitor on human glucose absorption after starch consumption. *Open Nutraceuticals J.* **2009**, *2*, 88–91. [CrossRef]
31. Onakpoya, I.; Aldaas, S.; Terry, R.; Ernst, E. The efficacy of Phaseolus vulgaris as a weight-loss supplement: A systematic review and meta-analysis of randomised clinical trials. *Br. J. Nutr.* **2011**, *106*, 196–202. [CrossRef] [PubMed]
32. Harvey, C.J.D.C.; Schofield, G.M.; Williden, M. The use of nutritional supplements to induce ketosis and reduce symptoms associated with keto-induction: A narrative review. *PeerJ* **2018**, *16*, e4488. [CrossRef] [PubMed]
33. Gardner, C.D.; Trepanowski, J.F.; Del Gobbo, L.C.; Hauser, M.E.; Rigdon, J.; Ioannidis, J.P.A.; Desai, M.; King, A.C. Effect of Low-Fat vs Low-Carbohydrate Diet on 12-Month Weight Loss in Overweight Adults and the Association With Genotype Pattern or Insulin Secretion. *JAMA* **2018**, *319*, 667–679. [CrossRef] [PubMed]

![foods logo] *foods*

MDPI

*Article*

# Phenolic Profiling and Antioxidant Capacity of *Eugenia uniflora* L. (Pitanga) Samples Collected in Different Uruguayan Locations

Ignacio Migues [1], Nieves Baenas [2], Amadeo Gironés-Vilaplana [2], María Verónica Cesio [1], Horacio Heinzen [1] and Diego A. Moreno [2],*

[1]   Facultad de Química, UdelaR, Av. Gral. Flores 2124, 11800 Montevideo, Uruguay; imigues@fq.edu.uy (I.M.); cs@fq.edu.uy (M.V.C.); heinzen@fq.edu.uy (H.H.)
[2]   CEBAS-CSIC, Food Science and Technology Department, Phytochemistry and Healthy Foods Laboratory, Campus Universitario Espinardo-25, E-30100 Espinardo, Murcia, Spain; nbaenas@cebas.csic.es (N.B.); amadeo.girones@elpozo.com (A.G.-V.)
*   Correspondence: dmoreno@cebas.csic.es; Tel.: +34-968-396200 (ext. 6369)

Received: 22 March 2018; Accepted: 20 April 2018; Published: 24 April 2018

**Abstract:** The use of nutrient-rich foods to enhance the wellness, health and lifestyle habits of consumers is globally encouraged. Native fruits are of great interest as they are grown and consumed locally and take part of the ethnobotanic knowledge of the population. Pitanga is an example of a native fruit from Uruguay, consumed as a jelly or an alcoholic beverage. Pitanga has a red-violet pigmentation, which is a common trait for foods that are a good source of antioxidants. Hence, fruits from different Uruguayan regions were analyzed via miniaturized sample preparation method, HPLC-DAD-ESI/MS$^n$ and RP-HPLC-DAD techniques to identify and quantify phenolic compounds, respectively. The antioxidant capacity was evaluated via DPPH and ORAC (Oxygen Radical Absorbance Capacity) assays. A multivariate linear regression was applied to correlate the observed antioxidant capacity with the phenolic content. Furthermore, Principal Components Analysis was performed to highlight characteristics between the various samples studied. The main results indicated differences between northern and southern Uruguayan samples. Delphinidin-3-hexoside was present in southern samples (mean of 293.16 µmol/100 g dry weight (DW)) and absent in the sample collected in the north (sample 3). All the samples contain high levels of cyanidin-3-hexoside, but a noticeable difference was found between the northern sample (150.45 µmol/100 g DW) and the southern sample (1121.98 µmol/100 g DW). The antioxidant capacity (mean ORAC of 56370 µmol Trolox®/100 g DW) were high in all the samples compared to the Food and Drug Administration (FDA) database of similar berry-fruits. The results of this study highlight the nutraceutical value of a native fruit that has not been exploited until now.

**Keywords:** *Eugenia uniflora* L.; nutraceuticals; antioxidant activity; polyphenols

## 1. Introduction

*Eugenia uniflora* L. belongs to Myrtaceae family and grows naturally in subtropical Latin-American zones [1]. Its cultivation has extended to other regions outside the American continent with similar climatic conditions. It grows mainly in Brazil, which is why it is known as "Brazilian Cherry", but it is also grown in Argentina, Paraguay and Uruguay [2]. Its fruits have a high carbohydrates content (around 38%) where maltose, lactose and fructose are the main identified compounds [3,4]. It has a high content of vitamin C, vitamin A, riboflavin (B12) and niacin (B3) [4]. Immature fruits show a high content of polyphenols that decreases with maturation [5]. On the other hand, carotenoid content increases with maturation evidenced by an increase in a reddish-orange coloration [6].

Pitanga leaves are often used in Brazilian traditional medicine due to its diuretic, antirheumatic, antifebrile, anti-inflammatory and hypocholesterolemic properties [7–10]. However, there is little information regarding the medicinal use of the fruit. The fruit has an acidic and sweet flavor and can be consumed fresh, in compotes, jams or juices [6]. Purple fleshed Pitanga fruit in its latest maturation stage has an edible portion (pulp and skin) of 61.76%, a vitamin C content of 38.35 mg/100 g and total anthocyanin content of 29.60 mg/100 g [11]. Additionally, a high content of total phenols (799.80 mg of gallic acid/100 g) and total carotenoids (5.86 µg of β-carotene/g) [4]. Recently, Pitanga juice showed an anti-inflammatory effect on oral gum epithelial cells; these results could be associated to the presence of cyanidin-3-glucoside and oxidoselina-1,3,7(11)-trien-8-one [12,13]. Hence, Pitanga's beneficial properties could be exploited in the nutraceutical industry.

According to DeFelice a nutraceutical can be defined as "a food or part of it, that has some health benefit, including the prevention and/or treatment of a disease" [14]. Therefore, research has driven evaluation/re-evaluation of foods and their beneficial properties. Antioxidant rich foods have received a lot of interest; promoting their consumption to decrease oxidative stress caused by stress, lack of sleep, poor diet, metabolic problems, etc.

Currently, a large number of different approaches to determine antioxidant activity have been established [15–17]. All tests differ in substrates, probes, reaction conditions, instrumentation, and quantification methods. Hence, it is difficult to compare the results obtained by one method or another. Based on the reactions involved, the tests can be further classified into two groups: HAT (Hydrogen Atom Transfer) and SET (Single Electron Transfer) [18]. In the HAT group, antioxidants must donate a hydrogen atom to stabilize the generated free radicals (a synthetic free radical generator and an oxidizable molecular probe are used to evaluate the kinetics of the reaction). SET-type assays involve a redox reaction where the antioxidant must donate an electron to the generated free radical. In both methods the "competition" with the oxidant radical is measured instead of the antioxidant capacity [18].

The ORAC (Oxygen Radical Absorbance Capacity) assay measures the overall antioxidant activity or capacity of a sample's ability to "quench or neutralize" peroxyl radicals. The peroxyl radicals are reactive species comparable to those ROS (Reactive Oxygen Species) biologically generated in the organism. In the ORAC assay, the peroxyl radicals, generated from the azo-compound AAPH (2,2′-azobis-(2-methylpropionamidine)dihydrochloride) react with fluorescein as a substrate. The fluorescence of the latter compound decreases over time, forming an area under the curve (fluorescence vs. time). In the presence of antioxidant compounds, the area under the curve increases linearly and proportionally to the concentration of antioxidants.

The ORAC assay quantifies via the HAT mechanism and measures the antioxidant activity of polyphenol and non-polyphenolic compounds present in each sample. It is important to note that the antioxidant activity does not have a direct correlation to the polyphenol nature of a sample [19]. The ORAC test reflects the overall capacity or antioxidant activity of a sample, due to the individual components and their additive, synergistic interactions. The ORAC value is usually expressed as micromoles of Trolox® equivalents/100 g of sample. Trolox® is an analogue of vitamin E and is often used as a comparative standard due to its solubility in water [18,20].

The DPPH method name stems from the reactant used (2,2-diphenyl-1-picrylhidracyl) to evaluate the ability of a sample's antioxidants to "quench or neutralize" a free radical. The DPPH assay utilizes molecules that differ completely from any free radical or reactive species generated by our organism as a source of free radicals. Although the reactant is easy to use, it places this method as a distant analytical approximation of the high reactivity that typically characterizes the ROS generated in biological systems [21]. This method is classified in the SET group. However, can be classified as a HAT-type assay according to the antioxidant present in the sample [13]. In this test the sample is incubated for 35 min and the decay of the absorbance is measured at 515 nm.

Based on previous reports of bioactivity [22] of South-Brazilian Pitanga fruit, the objective of the present work was to extract, identify and quantify the phenolic compounds in mature fruits of *Eugenia uniflora* L. The antioxidant capacity was evaluated via ORAC and DPPH; and the phenolic profiles of *E. uniflora* from different Uruguayan locations were compared. Pitanga is considered a South American native fruit and its consumption is not among the most common in the region. However, its ethnobotanical characteristics, traditional uses and its antioxidant properties highlights its potential to improve the quality of life of those who consume it [23]. Hence, this study seeks to explore and further promote the value of the Uruguayan native Pitanga fruit that has not been exploited until now.

## 2. Materials and Methods

### 2.1. Chemicals

2,2-Diphenyl-1-picrylhidracyl (DPPH), fluorescein, 2,2′-azobis-(2-methylpropionamidine) dihydrochloride (AAPH), monobasic sodium phosphate and dibasic sodium phosphate were obtained from Sigma-Aldrich (Steinheim, Germany). 6-hydroxy-2,5,7,8-tetramethylchroman-2-carboxilic acid (Trolox®) was purchased from Fluka Chemika (Neu-Ulm, Germany). Cyanidin-3-*O*-glucoside and rutin were obtained from Polyphenols Laboratories AS (Sandnes, Norway). Ultrapure water was produced using a Millipore Milli-Q® Ultrapure Water Solutions Type 1. All solvents used were HPLC grade from Sigma-Aldrich (Steinheim, Germany).

### 2.2. Samples

The fruits (purple fleshed Pitanga breeding lines) were collected in different Uruguayan locations (described below), they were identified and kept at the Jose Arechavaleta Herbarium in the Faculty of Chemistry, UdelaR, Uruguay (Voucher number MVFQ 4427). Samples 1, 5, 6 & 7 were collected in the north of Montevideo Department (−34.804951, −56.230206) in December 2014, November 2015, December 2015 and April 2016, respectively. Sample 2 was collected in the south of Montevideo Department (−34.884536, −56.073039), sample 3 in Paysandú Department (−32.322604, −58.088243) and sample 4 in Ciudad de la Costa, Canelones Department (−34.799105, −55.908381); samples 2, 3 & 4 were collected in December 2014. All samples were freeze-dried, and the water content calculated by weight before and after lyophilization.

### 2.3. Extraction

Each lyophilized grinded sample (100 mg) was mixed with 1 mL of methanol/water/formic acid (70:29:1, $v/v/v$) in a 2 mL conical polypropylene tube (Eppendorf, Madrid, Spain). Then, the samples were vortexed and subjected to indirect sonication in an ultrasound cleaning bath for 60 min (BRANSONIC® Ultrasonic cleaner mod. 5510E-MTH, Ultrasonic frequency: 135 W or 42 KHz ± 6%). The samples were kept overnight at 4 °C and sonicated again for 60 min. The supernatant was separated from the solid residue after centrifugation (9500× $g$, 15 min), filtered using a 0.45 μm PVDF filter (Millex HV13, Millipore, Bedford, MA, USA) and stored at 4 °C until they were analyzed.

### 2.4. Identification of Phenolic Compounds Via HPLC-DAD-ESI/MS$^n$ and Quantification Via RP-HPLC-DAD

The identification analyses were carried out using an Agilent HPLC 1100 series model equipped with a photodiode array detector and a mass detector in series (Agilent Technologies, Waldbronn, Germany). It consisted of a binary pump (model G1312A), a degasser (model G1322A), an autosampler (model G1313A), and a photodiode array detector (model G1315B). The HPLC system was controlled by ChemStation for LC 3D Systems software Rev. B.01.03-SR2 (204) (Agilent Technologies Spain S.L., Madrid, Spain). The mass detector was an ion trap spectrometer (model G2445A) equipped with an electrospray ionization interface and was controlled by LC/MS software (Esquire Control Ver. 6.1. Build No. 534.1., Bruker Daltoniks GmbH, Bremen, Germany). The ionization conditions were 350 °C capillary temperature and 4 kV voltage, the nebulizer pressure was 65.0 psi and the nitrogen flow rate was 11 L/min. Full-Scan mass covered the range of $m/z$ from 100 to 1200. Collision-induced fragmentation experiments were performed in the ion trap using helium as the collision gas, with voltage ramping cycles from 0.3 to 2 V. The mass spectrometry data were acquired in the positive ionization mode for anthocyanins and in the negative ionization mode for other flavonoids. The $MS^n$ was carried out in the automatic mode on the more abundant fragment ion in $MS^{(n-1)}$. A Luna $C_{18}$ column (250 × 4.6 mm, 5 µm particle diameter; Phenomenex, Macclesfield, UK) was used. Mobile phase A: water/formic acid (99:1, $v/v$), mobile phase B: acetonitrile, initial conditions: 8% solvent B, reaching 15% solvent B at 25 min, 22% at 55 min, and 40% at 60 min, which was maintained isocratic until 70 min. The flow rate was 0.8 mL/min during the whole run, all gradients were linear, and the injection volume was 7 µL. Chromatograms were recorded at 280, 320, 360 and 520 nm.

For quantification experiments the same conditions were applied, except for the injection volume was set to 20 µL and the flow rate was 0.9 mL/min. Anthocyanins were quantified as cyanidin 3-*O*-glucoside at 520 nm and flavonols as rutin at 360 nm.

### 2.5. DPPH Antioxidant Activity Measurements

Experiments were carried out using 96-well micro plates (Nunc, Roskilde, Denmark) and an Infinite® M200 micro plate reader (Tecan, Grödig, Austria). The antioxidant activity was evaluated measuring the change of the absorbance at 515 nm after 35 min of reaction with the radical DPPH· (2 µL of the sample + 250 µL of DPPH· solution). The results were expressed as µmol of Trolox®/100 g dry weight.

### 2.6. ORAC Antioxidant Activity Measurements

The antioxidant assay was performed using black-walled 96-well plates (Nunc, Roskilde, Denmark) and an Infinite® M200 micro plate reader (Tecan, Grödig, Austria). Each well with a final volume of 200 µL. 10 mM phosphate buffer (pH 7.4) was used to prepare 1 µM fluorescein and 250 mM AAPH solutions. Each well received 150 µL of fluorescein solution and 25 µL of phosphate buffer, Trolox® solutions or sample solution to measure the blank, the curve or the samples respectively. The plate was placed into the microplate reader and after 30 min of incubation at 37 °C, 25 µL AAPH solution were added to each well and fluorescence was recorded every 5 min for 120 min using an excitation wavelength of 485 nm and an emission wavelength of 520 nm. ORAC values were calculated using the difference in Areas Under the Fluorescein Decay Curve (AUC) between the blank and a sample. The results were expressed as µmol of Trolox®/100 g dry weight.

### 2.7. Statistical Analysis

Data shown are mean values ($n = 3$), subjected to Analysis of Variance (ANOVA) and multiple range test (Tukey's test), using RStudio software (Version 1.1.383, RStudio, Boston, MA, USA) and InfoStat (Version 2017, Universidad Nacional de Córdoba, Córdoba, Argentina) [24].

## 3. Results and Discussion

### 3.1. Water Content

The results of the water content of the samples are shown in Table 1.

**Table 1.** Water content calculated by weighing the samples before and after lyophilization.

| Sample | Location/Sampling Date | % Water (% $w/w$) [1] |
|:------:|:----------------------:|:---------------------:|
| 1 | North of Montevideo/December 2014 | 68.1 [c,d] ± 0.9 |
| 2 | South of Montevideo/December 2014 | 63.6 [b,c] ± 1.7 |
| 3 | Paysandú/December 2014 | 66.1 [b,c,d] ± 0.9 |
| 4 | Canelones/December 2014 | 46.4 [a] ± 6.4 |
| 5 | North of Montevideo/November 2015 | 66.5 [b,c,d] ± 4.1 |
| 6 | North of Montevideo/December 2015 | 58.3 [b] ± 0.6 |
| 7 | North of Montevideo/April 2016 | 73.0 [d] ± 2.7 |

[1] Means ($n = 3$) followed by different letters are significantly different at $p < 0.05$ according to Tukey's test.

The stress generated by the lack of precipitations, is associated to the increased concentration of sugar and phenolic compounds in grapes [25]. For grapes this enhances fermentation and wine quality. However, in a study where water availability and photosynthetic efficiency of Pitanga plants was evaluated the main influencing factor that determined the quality of the fruits was not water availability, but light exposure [26].

During 2014, Uruguay suffered from excess precipitations, especially in the south of the country (60% above the regular precipitation rate) [27]. However, sample 4 showed a smaller water %, which could be due to its location near the beach, where water availability was less due to the sandy ground. During 2015, Uruguay suffered from slight droughts and higher temperatures at the end of the year, which could be the cause of the decreased water content in sample 6. The first two months of 2016 experienced a lot of rain and consequently, the water content of sample 7 was much higher in comparison to the other studied samples.

## 3.2. Identification and Quantification of Phenolic Compounds

As expected, all samples resulted in similar phenolic profiles. Hence, the different climate situations had no influence, see Table 2. Two anthocyanins were identified, delphinidin-3-hexoside and cyanidin-3-hexoside, the latter was the main anthyocyanin present in all samples. The results are in agreement with previously obtained values by Einbond in 2004 [28]. Thirteen flavonols were identified and myricetin-rhamnoside was the most abundant in all samples. Samples collected from northern Montevideo showed variable levels of total flavonol content (TFC) that ranged from 120 to 860 µmol/100 g dry weight, but the Paysandú sample showed a much higher level of TFC (1050 µmol/100 g dry weight), see Table 3.

The total anthocyanin content (TAC), total quercetin derivatives content (TQC), total myricetin derivatives content (TMC), total flavonol content (TFC), total catechol derivatives content (TCC), total pyrogallol derivatives content (TPC) and total polyphenol content (TPPC) were calculated for each sample and the results are shown in Table 3.

## 3.3. Antioxidant Activity

The results of the antioxidant assays (ORAC and DPPH) are shown in Table 4, the results for both assays are expressed as (µmol Trolox®/100 g dry weight).

In general, samples collected in the north of Montevideo showed greater level of ORAC scavenging activity, but similar levels of DPPH activity compared to the other samples. The direct correlation between ORAC-TAC, ORAC-TFC, DPPH-TAC, and DPPH-TFC was not statistically significant ($R^2 < 0.3$ in all cases). However, the ORAC and DPPH activity of all Pitanga samples were higher compared to other Latin-American berries [29].

All the samples collected in southern Uruguay contained delphinidin together with cyanidin. This is a clear difference with the findings of Celli [5] that described cyanidin as the only occurring anthocyanin in Brazilian Pitanga samples. Northern Uruguay is naturally connected with southern Brazil but the continuity towards southern Uruguay is broken by the Negro river, these results could point out a region related chemo-diversity. Furthermore, three other myricetin derivatives were not identified in the Brazilian samples.

Studying the linear correlation (Figure 1) between the identified compounds grouped as it is shown in Table 3, we can see that when the concentration of anthocyanins increased, the flavonol content decreased, that is a negative correlation (shown in red spots, bigger spots represents a stronger correlation). A strong positive correlation between quercetin and myricetin derivatives, and a negative correlation of both flavonol groups to the anthocyanins concentration is shown. When the biosynthesis of anthocyanins is favored, there is less biosynthesis of flavonols as expected from their common biosynthetic pathway, similarly described for bilberries [30].

**Table 2.** Phenolic compounds identified and quantified (μmol/100 g dry weight) in different samples [1].

| Compound [2] | Rt | [M − H]+ | MSn | Sample 1 | Sample 2 | Sample 3 | Sample 4 | Sample 5 | Sample 6 | Sample 7 |
|---|---|---|---|---|---|---|---|---|---|---|
| C1 | 29.8 | 465 | 303 | 19.99 a ± 3.34 | 373.24 c ± 3.98 | ND 3,a | 381.38 c ± 31.81 | 282.99 b ± 24.49 | 215.53 b ± 19.42 | 485.81 d ± 26.25 |
| C2 | 33.8 | 449 | 287 | 115.01 a ± 8.56 | 1757.29 d ± 18.34 | 150.45 a ± 0.25 | 987.77 c ± 49.63 | 916.35 b,c ± 85.15 | 706.78 b ± 68.87 | 2248.68 e ± 98.76 |
| | | [M − H]− | | | | | | | | |
| C3 | 48.4 | 631 | 479, 317, 271 | 8.07 d ± 1.15 | 14.18 f ± 0.34 | 10.67 e ± 0.34 | ND 3,a | 3.63 b,c ± 0.08 | 2.41 b ± 0.57 | 4.94 c ± 0.36 |
| C4 | 53.6 | 479 | 317, 271 | 34.33 c ± 1.28 | 46.41 d ± 0.43 | 56.53 e ± 1.11 | 6.11 a ± 0.18 | 12.97 b ± 1.52 | 8.77 a ± 0.86 | 17.05 b ± 2.14 |
| C5 | 54.7 | 479 | 317, 271 | 32.37 d ± 2.52 | 35.82 d ± 0.34 | 22.59 c ± 1.49 | 6.53 a,b ± 0.50 | 4.85 a ± 1.76 | 7.57 a,b ± 0.78 | 10.25 b ± 2.45 |
| C6 | 56.8 | 449 | 317, 271, 179 | 23.44 d ± 0.26 | 6.79 b,c ± 0.51 | 9.39 c ± 4.91 | 1.22 a ± 1.46 | 2.15 a,b ± 0.24 | 1.52 a ± 0.18 | 3.33 a,b ± 0.35 |
| C7 | 58.0 | 615 | 463, 301, 271 | 8.45 c ± 0.09 | 11.10 d ± 0.64 | 16.87 e ± 0.43 | 0.92 a ± 0.81 | 7.61 b,c ± 0.65 | 5.27 b ± 0.75 | 11.59 d ± 1.19 |
| C8 | 61.0 | 449 | 317, 271, 179 | 49.91 d ± 3.46 | 4.57 a,b ± 0.94 | 26.30 c ± 0.85 | 2.09 a ± 0.20 | 5.33 a,b ± 1.10 | 3.96 a ± 0.78 | 8.34 b ± 1.11 |
| C9 | 61.9 | 463 | 317, 271, 179 | 378.69 f ± 10.72 | 138.43 d ± 3.37 | 243.25 e ± 1.20 | 16.81 a ± 1.04 | 41.63 b ± 3.49 | 29.16 a,b ± 3.76 | 56.62 c ± 3.33 |
| C10 | 62.9 | 463 | 301 | 29.25 c ± 2.13 | 64.65 d ± 2.39 | 167.25 e ± 0.00 | 4.11 a ± 1.50 | 14.97 b ± 0.59 | 8.17 a ± 0.66 | 18.48 b ± 1.05 |
| C11 | 63.5 | 463 | 301 | 38.30 b ± 2.65 | 71.26 c ± 4.27 | 101.54 d ± 3.59 | 13.11 a ± 3.81 | 8.22 a ± 0.80 | 8.32 a ± 1.02 | 10.61 a ± 0.20 |
| C12 | 64.5 | 433 | 301 | 8.97 c ± 1.24 | 10.59 c ± 0.04 | 19.64 d ± 0.30 | 2.57 a,b ± 0.14 | 2.41 a,b ± 0.48 | 1.75 a ± 0.42 | 4.17 b ± 1.35 |
| C13 | 64.9 | 433 | 301 | 28.44 d ± 1.75 | 8.71 c ± 2.43 | 55.81 e ± 0.38 | 6.55 a ± 1.35 | 11.36 a,b ± 2.30 | 6.79 a ± 0.96 | 16.48 b,c ± 1.55 |
| C14 | 65.4 | 433 | 301 | 36.12 d ± 0.43 | 36.21 d ± 0.13 | 92.95 e ± 0.17 | 3.37 a ± 0.51 | 15.12 b ± 2.32 | 9.98 b ± 2.93 | 22.92 c ± 2.90 |
| C15 | 65.7 | 447 | 301 | 80.74 d ± 0.13 | 93.13 e ± 1.54 | 224.17 f ± 0.04 | 15.21 a ± 0.33 | 34.62 b ± 3.82 | 26.44 b ± 4.27 | 53.22 c ± 3.34 |

[1] Means (n = 3) in the same rows followed by different letters are significantly different at $p < 0.05$ according to Tukey's test. [2] Compounds: anthocyanins quantified at 520 nm: C1: delphinidin-3-hexoside, C2: cyanidin-3-hexoside. Flavonols quantified at 360 nm: C3: myricetin-galloyl-hexoside, C4: myricetin-hexoside 1, C5: myricetin-hexoside 2, C6: myricetin-pentoside 1, C7: quercetin galloyl hexoside, C8: myricetin-pentoside 2, C9: myricetin-rhamnoside, C10: quercetin hexoside 1, C11: quercetin hexoside 2, C12: quercetin pentoside 1, C13: quercetin pentoside 2, C14: quercetin pentoside 3 and C15: quercetin rhamnoside. [3] ND = not detected.

**Table 3.** Total anthocyanin content (TAC), total quercetin derivatives content (TQC), total myricetin derivatives content (TMC), total flavonol content (TFC), total catechol derivatives content (TCC), total pyrogallol derivatives content (TPC) and total polyphenol content (TPPC) results expressed as μmol/100 g dry weight.

| | Sample 1 | Sample 2 | Sample 3 | Sample 4 | Sample 5 | Sample 6 | Sample 7 |
|---|---|---|---|---|---|---|---|
| TAC | 134.98 a ± 11.90 | 2130.53 d ± 14.35 | 150.45 a ± 0.24 | 1369.15 c ± 79.26 | 1199.35 b,c ± 109.47 | 922.32 b ± 87.78 | 2734.50 e ± 125.00 |
| TQC | 230.26 d ± 1.61 | 308.45 e ± 8.03 | 678.25 f ± 4.14 | 45.84 a ± 2.98 | 94.33 b ± 8.56 | 66.70 a ± 7.42 | 137.44 c ± 11.21 |
| TMC | 526.83 f ± 17.08 | 246.20 d ± 3.43 | 368.79 e ± 6.24 | 32.76 a ± 1.65 | 70.54 b ± 8.14 | 53.39 a,b ± 6.56 | 100.50 c ± 9.57 |
| TFC | 757.09 f ± 18.69 | 554.65 e ± 11.47 | 1047.04 g ± 10.38 | 78.60 a ± 2.50 | 164.87 c ± 14.57 | 120.08 b ± 7.39 | 237.94 d ± 17.48 |
| TCC | 345.25 a ± 10.18 | 2065.75 d ± 10.30 | 828.70 b,c ± 4.38 | 1033.62 c ± 49.24 | 1010.68 c ± 92.65 | 773.48 b ± 65.14 | 2386.12 e ± 100.13 |
| TPC | 546.82 c ± 20.42 | 619.43 c ± 7.42 | 368.79 b ± 6.28 | 414.13 b ± 31.28 | 353.54 a,b ± 32.23 | 268.92 a ± 25.93 | 586.32 c ± 35.30 |
| TPPC | 892.07 a ± 30.59 | 2685.18 d ± 2.89 | 1197.49 b,c ± 10.62 | 1447.75 c ± 77.27 | 1364.21 c ± 124.02 | 1042.40 a,b ± 90.94 | 2972.43 d ± 135.41 |

Means (n = 3) in the same rows followed by different letters are significantly different at $p < 0.05$ according to Tukey's test.

**Table 4.** Oxygen Radical Absorbance Capacity (ORAC) and 2,2-diphenyl-1-picrylhidracyl (DPPH) antioxidant activity results for each sample.

| Assay | Sample 1 | Sample 2 | Sample 3 | Sample 4 | Sample 5 | Sample 6 | Sample 7 |
|---|---|---|---|---|---|---|---|
| ORAC | 57,440 [b,c] ± 1090 | 44,510 [a,b] ± 460 | 22,800 [a] ± 300 | 57,560 [b,c] ± 10,040 | 57,400 [b,c] ± 16,380 | 82,340 [c] ± 9070 | 72,550 [c] ± 15,280 |
| DPPH | 44,170 [b] ± 4480 | 11,940 [a] ± 2200 | 10,070 [a] ± 1200 | 12,200 [a] ± 650 | 15,430 [a] ± 70 | 12,960 [a] ± 1250 | 13,950 [a] ± 560 |

Means ($n = 3$) in the same rows followed by different letters are significantly different at $p < 0.05$ according to Tukey's test.

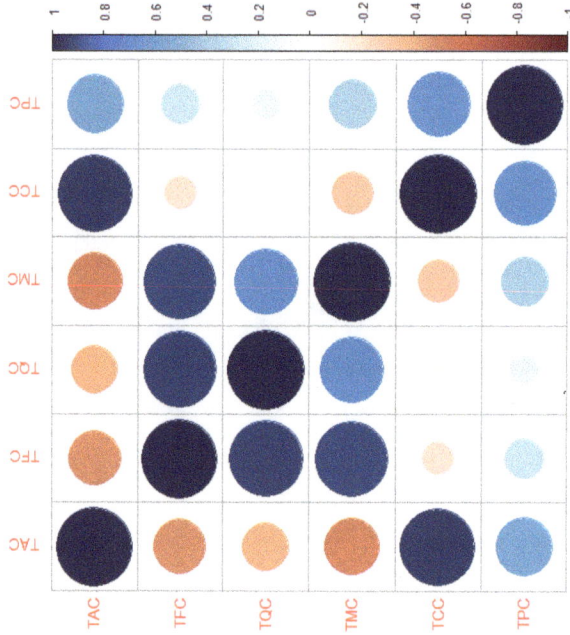

**Figure 1.** Correlation boxplot between identified and quantified compounds.

A PCA was performed using the phenolic content to identify the compounds that weighted the most when classifying the samples. Figure 2a shows that all the samples are grouped together according mainly to PC1 (69.62% of the variance), except for the sample collected in Paysandú (sample 3). Sample 7 is located closer to the anthocyanins than the rest of the samples, mainly because of PC2 (15.68% of the variance).

When we analyze the samples collected in 2014, sample 4 (Canelones) and 2 (south of Montevideo) correlate with a high anthocyanin content, which is coherent with the TAC of the samples. On the other hand, Sample 1 (north of Montevideo) correlated with the myricetin glycosides (C6, C8 and C9), while sample 3 had a stronger correlation to the quercetin derivatives. This PCA diagram explains 88.11% of the total variance (PC1 = 68.58% and PC2 = 19.53%).

Samples 1, 4, 5, 6 and 7 were labeled as high ORAC samples (H), while samples 2 and 3 were labeled as low ORAC samples (L). The compounds that classified the H samples were the anthocyanins while the flavonols classified the L samples, see Figure 3.

In order to predict the ORAC and DPPH values of a new sample based on the concentration of the identified compounds, we performed a stepwise multivariate linear regression analysis, the equations described below were obtained:

$$ORAC = 69825.66 - 552.22[C11] + 378.91[C5]$$
$$(R^2 = 0.7345, p\text{-value} = 1.876 \times 10^{-5}, RSE = 9050 \text{ on } 15 \text{ Degrees of Freedom})$$

(1)

$$DPPH = 14108.99 - 97.55[C11] + 168.64[C9] - 246.19[C15] + 421.48[C14] - 260.30[C5] -$$
$$367.10[C8] - 80.43[C10] + 287.09[C4] - 527.05[C3] - 80.35[C6]$$
$$(R^2 = 0.9995, p\text{-value} = 8.487 \times 10^{-12}, RSE = 238.3 \text{ on } 7 \text{ Degrees of Freedom})$$

(2)

An increased number of reports have shown the non-additivity of the antioxidant properties of polyphenols, in the ORAC and DPPH assays. Particularly, they can act negatively on the overall value of the measurement, depending on the certain mixture of polyphenols and the point they enter in the reaction chain cycle, either as oxidants or reductants. The anthocyanins, due to their positive formal charge, are kinetically more reactive than the neutral polyphenols when reacting with DPPH. The reaction chain thus triggered could foster negative contributions of the quercetin derivatives, but positive ones from the myricetin ones, based on the positive or negative contributions shown in Equation (2). These mixture effects were "partly explained by regeneration mechanisms between antioxidants, depending on the chemical structure of molecules and on the possible formation of stable intermolecular complexes" [31]. Further work is required to understand the global behavior of complex mixtures of different classes of polyphenols in these widely used bench-top bioassays.

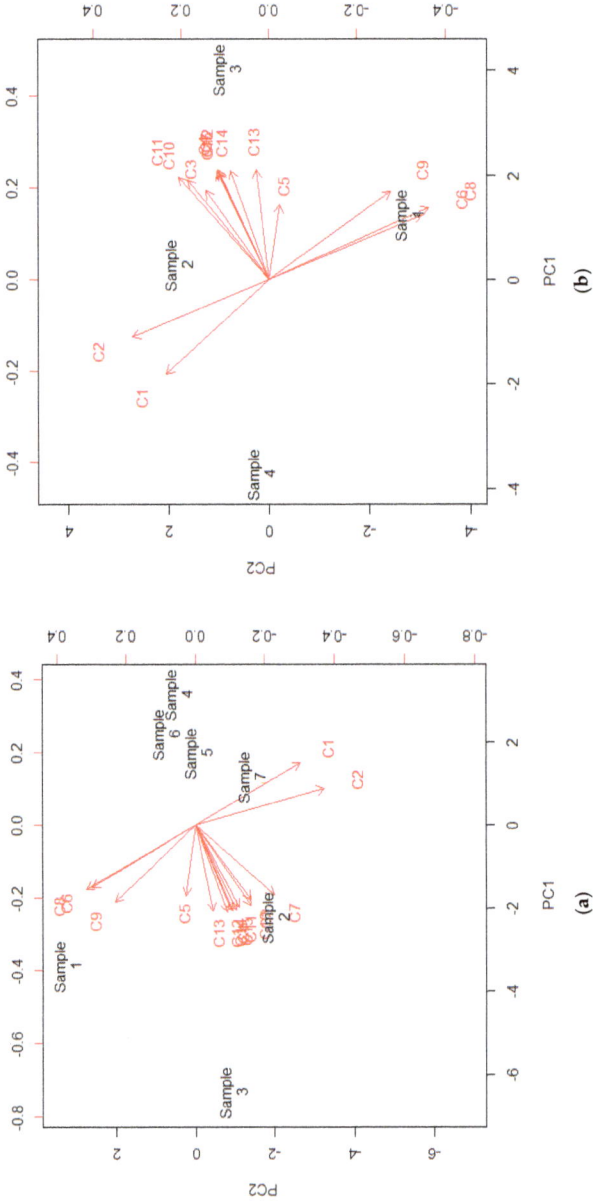

**Figure 2.** Phenolic content PCA of: (**a**) all samples, and (**b**) samples collected in December 2014.

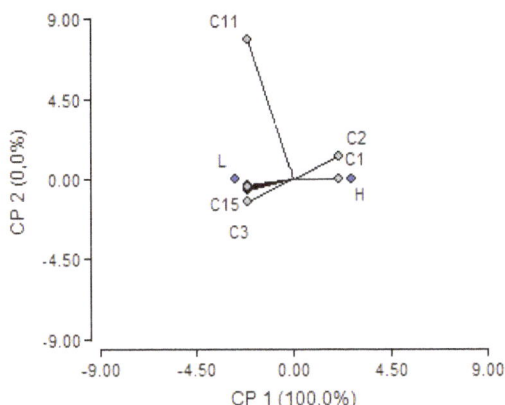

**Figure 3.** PCA where all identified compounds are classified according to low (L) and high (H) ORAC results.

## 4. Conclusions

The samples from different Uruguayan locations were very similar on their composition but presented different levels of antioxidant activity. The sample collected in 2014 in the north of Montevideo showed the highest ORAC and DPPH values. Several samples were collected until April 2016 and the evaluated. There is no clear relationship between the TAC and TFC values and the bioactivity studied; due to the presence of secondary metabolites with antioxidant activity in Pitanga fruit, for example carotenoids [6]. However, we could identify the phenolic compounds responsible for the separation between samples using PCA diagrams. Furthermore, the different location samples were characterized by specific types of phenolic compounds.

The sample collected in April 2016 (sample 7) showed a TPC three times higher than the sample collected in December 2015 (sample 6) probably because of the ripening during summer season where light exposure enhances the quality of the fruits.

In contrast to the results obtained by Celli et al. [5], the main compounds identified amongst all flavonols were myricetin derivatives and not quercetin derivatives, these results can be used as a characteristic of Uruguayan Pitanga fruits, with high antioxidant capacity.

**Author Contributions:** D.A.M. conceived and designed the experiments; I.M., A.G.-V. and N.B. performed the experiments; I.M., A.G.-V. and N.B. analyzed the data; D.A.M. contributed reagents/materials/analysis tools; I.M., M.V.C. and H.H. gathered the samples and wrote the drafts as well as the final version and the corrections of the paper.

**Acknowledgments:** The authors gratefully acknowledge the CYTED program for the support of the international collaboration through the Thematic Network AGL 112RT0460 CORNUCOPIA.

**Conflicts of Interest:** The authors declare no conflict of interest.

## References

1. Bicas, J.L.; Molina, G.; Dionísio, A.P.; Barros, F.F.C.; Wagner, R.; Maróstica, M.R.; Pastore, G.M. Volatile constituents of exotic fruits from Brazil. *Food Res. Int.* **2011**, *44*, 1843–1855. [CrossRef]
2. Consolini, A.E.; Sarubbio, M.G. Pharmacological effects of Eugenia uniflora (Myrtaceae) aqueous crude extract on rat's heart. *J. Ethnopharmacol.* **2002**, *81*, 57–63. [CrossRef]
3. Amoo, A.; Adebayo, O.; Oyeleye, A. Chemical evaluation of winged beans (*Psophocarpus Tetragonolobus*), Pitanga cherries (*Eugenia uniflora*) and orchid fruit (*Orchid fruit myristica*). *Afr. J. Food Agric. Nutr. Dev.* **2006**, *6*, 3–12.

4.  Vasconcelos Costa, A.G.; Garcia-Diaz, D.F.; Jimenez, P.; Ibrahim Silva, P. Bioactive compounds and health benefits of exotic tropical red–blackberries. *J. Funct. Foods* **2013**, *5*, 539–549. [CrossRef]
5.  Celli, G.B.; Pereira-Netto, A.B.; Beta, T. Comparative analysis of total phenolic content, antioxidant activity, and flavonoids profile of fruits from two varieties of Brazilian cherry (*Eugenia uniflora* L.) throughout the fruit developmental stages. *Food Res. Int.* **2011**, *44*, 2442–2451. [CrossRef]
6.  Galvão de Lima, V.L.; Almeida Mélo, E.; da Silva Lima, D.E. Fenólicos e carotenóides totais em pitanga. *Sci. Agric.* **2002**, *59*, 447–450. [CrossRef]
7.  Corrêa, P. *Dicionário das Plantas Uteis do Brasil e das Exóticas Cultivadas*; Imprensa National-Ministerio da Agricultural: Rio de Janeiro, Brazil, 1984; Volume 5.
8.  Rücker, G.; Brasil e Silva, G.A.A.; Bauer, L.; Schikarski, M. Neu Inhaltstoffe von *Stenoealyx michelii* (New constituents of Stenoealyx michelii). *Planta Med.* **1977**, *31*, 322–327. [CrossRef]
9.  Ferro, E.; Schinini, A.; Maldonado, M.; Rosner, J.; Schmeda-Hirschmann, G. *Eugenia unliflora* leaf extract and lipid metabolism in Cebus apella monkeys. *J. Ethnopharmacol.* **1988**, *24*, 321–325. [CrossRef]
10. Schapoval, E.E.S.; Silveira, S.M.; Miranda, M.L.; Alice, C.B.; Henriques, A.T. Evaluation of some pharmacological activities of *Eugenia uniflora* L. *J. Ethnopharmacol.* **1994**, *44*, 137–142. [CrossRef]
11. Vizzotto, M.; Cabral, L.; Santos Lopes, S. Pitanga (*Eugenia uniflora* L.). In *Postharvest Biology and Technology of Tropical and Subtropical Fruits. Mangosteen to White Sapote*; Yahia, E., Ed.; Woodhead Publishing: Cambridge, UK, 2011; Volume 4, pp. 272–286.
12. Soares, D.J.; Walker, J.; Pignitter, M.; Walker, J.M.; Imboeck, J.M.; Ehrnhoefer-Ressler, M.M.; Montenegro Brasil, I.; Veronika, S. Pitanga (*Eugenia uniflora* L.) fruit juice and two major constituents thereof exhibit anti-inflammatory properties in human gingival and oral gum epithelial cells. *Food Funct.* **2014**, *5*, 2981–2988. [CrossRef] [PubMed]
13. Soares, D.J.; Pignitter, M.; Ehrnhöfer-Ressler, M.M.; Walker, J.; Montenegro, I.; Somoza, V. Identification and Quantification of Oxidoselina-1,3,7-Trien-8-One and Cyanidin-3-Glucoside as One of the Major Volatile and Non-Volatile Low-Molecular- Weight Constituents in Pitanga Pulp. *PLoS ONE* **2015**, *10*, e0138809. [CrossRef]
14. DeFelice, S.L. The nutraceutical revolution: Its impact on food industry R&D. *Trends Food Sci. Technol.* **1995**, *6*, 59–61.
15. Tai, A.; Sawano, T.; Yazama, F.; Ito, H. Evaluation of antioxidant activity of vanillin by using multiple antioxidant assays. *Biochim. Biophys. Acta Gen. Subj.* **2011**, *1810*, 170–177. [CrossRef] [PubMed]
16. Jones, A.; Acquaviva, A.; Dennis, G.R.; Shalliker, R.A.; Soliven, A. High throughput screening of phenolic constituents in a complex sample matrix using post-column derivatisations employing reaction flow HPLC columns. *Microchem. J.* **2018**. [CrossRef]
17. Acquaviva, A.; Jones, A.; Dennis, G.R.; Shalliker, R.A.; Soliven, A. Phenolic profiling of complex tea samples via simultaneous multiplexed detection employing reaction flow HPLC columns and colorimetric post column derivatisation. *Microchem. J.* **2018**, *138*, 533–539. [CrossRef]
18. Huang, D.; Ou, B.; Prior, R.L. The Chemistry behind Antioxidant Capacity Assays. *J. Agric. Food Chem.* **2005**, *53*, 1841–1856. [CrossRef] [PubMed]
19. Acquaviva, A.; Jones, A.; Dennis, G.R.; Shalliker, R.A.; Soliven, A. Detection selectivity in the analysis of "reactive" chemical compounds derived from natural samples via reaction flow chromatography. *Microchem. J.* **2018**, *139*, 315–321. [CrossRef]
20. Prior, R.L.; Wu, X.; Schaich, K. Standardized Methods for the Determination of Antioxidant Capacity and Phenolics in Foods and Dietary Supplements. *J. Agric. Food Chem.* **2005**, *53*, 4290–4302. [CrossRef] [PubMed]
21. Xie, J.; Schaich, K.M. Re-evaluation of the 2,2-diphenyl-1-picrylhydrazyl Free Radical (DPPH) Assay for Antioxidant Activity. *J. Agric. Food Chem.* **2014**, *62*, 4251–4260. [CrossRef] [PubMed]
22. Bagetti, M.; Facco, E.M.P.; Piccolo, J.; Hirsch, G.E.; Rodriguez-Amaya, D.; Kobori, C.N.; Vizzotto, M.; Emanuelli, T. Physicochemical characterization and antioxidant capacity of pitanga fruits (*Eugenia uniflora* L.). *Ciênc. Tecnol. Aliment.* **2011**, *31*, 147–154. [CrossRef]
23. Lim, T.K. *Edible Medicinal and Non-Medicinal Plants*; Springer: Dordrecht, The Netherlands, 2012; pp. 620–630, ISBN 978-94-007-1764-0.
24. Di Rienzo, J.; Casanoves, F.; Balzarini, M.G.; Gonzalez, L.; Tablada, M.; Robledo, C.W. *InfoStat Versión 2017*; InfoStat: Córdoba, Argentina, 2017.

25. Romero, P.; Gil-Muñoz, R.; del Amor, F.M.; Valdés, E.; Fernández, J.I.; Martinez-Cutillas, A. Regulated Deficit Irrigation based upon optimum water status improves phenolic composition in Monastrell grapes and wines. *Agric. Water Manag.* **2013**, *121*, 85–101. [CrossRef]

26. Mielke, M.S.; Schaffer, B. Photosynthetic and growth responses of *Eugenia uniflora* L. seedlings to soil flooding and light intensity. *Environ. Exp. Bot.* **2010**, *68*, 113–121. [CrossRef]

27. INUMET ANOMALIAS CLIMATICAS Durante 2014 en Uruguay. Available online: http://www.meteorologia.com.uy/reportes/noticias/docs/pdf/rrpp/168_53868.pdf (accessed on 15 October 2016).

28. Einbond, L.S.; Reynertson, K.A.; Luo, X.-D.; Basile, M.J.; Kennelly, E.J. Anthocyanin antioxidants from edible fruits. *Food Chem.* **2004**, *84*, 23–28. [CrossRef]

29. Gironés-Vilaplana, A.; Baenas, N.; Villaño, D.; Speisky, H.; García-Viguera, C.; Moreno, D.A. Evaluation of Latin-American fruits rich in phytochemicals with biological effects. *J. Funct. Foods* **2014**, *7*, 599–608. [CrossRef]

30. Jaakola, L.; Määttä, K.; Pirttilä, A.M.; Törrönen, R.; Kärenlampi, S.; Hohtola, A. Expression of Genes Involved in Anthocyanin Biosynthesis in Relation to Anthocyanin, Proanthocyanidin, and Flavonol Levels during Bilberry Fruit Development. *Plant Physiol.* **2002**, *130*, 729–739. [CrossRef] [PubMed]

31. Peyrat-Maillard, M.N.; Cuvelier, M.E.; Berset, C. Antioxidant activity of phenolic compounds in 2,2′-azobis (2-amidinopropane) dihydrochloride (AAPH)-induced oxidation: Synergistic and antagonistic effects. *J. Am. Oil Chem. Soc.* **2003**, *80*, 1007–1012. [CrossRef]

![foods logo] *foods*

MDPI

Review

# Nanoparticles and Controlled Delivery for Bioactive Compounds: Outlining Challenges for New "Smart-Foods" for Health

MCarment Martínez-Ballesta [1], Ángel Gil-Izquierdo [2], Cristina García-Viguera [2] and Raúl Domínguez-Perles [2,*]

[1]  Department of Plant Nutrition, Centro de Edafología y Biología Aplicada del Segura-Spanish Council for Scientific Research (CEBAS-CSIC), Campus de Espinardo 25, 30100 Espinardo, Murcia, Spain; mballesta@cebas.csic.es

[2]  Research Group on Quality, Safety and Bioactivity of Plant Foods, Department of Food Science and Technology, Centro de Edafología y Biología Aplicada del Segura-Spanish Council for Scientific Research (CEBAS-CSIC), Campus de Espinardo 25, 30100 Espinardo, Murcia, Spain; angelgil@cebas.csic.es (A.G.-I.); cgviguera@cebas.csic.es (C.G.-V.)

*   Correspondence: rdperles@cebas.csic.com or rdperles@cebas.csic.es; Tel.: +34-968-396-200 (ext. 6247)

Received: 21 March 2018; Accepted: 4 May 2018; Published: 7 May 2018

**Abstract:** Nanotechnology is a field of research that has been stressed as a very valuable approach for the prevention and treatment of different human health disorders. This has been stressed as a delivery system for the therapeutic fight against an array of pathophysiological situations. Actually, industry has applied this technology in the search for new oral delivery alternatives obtained upon the modification of the solubility properties of bioactive compounds. Significant works have been made in the last years for testing the input that nanomaterials and nanoparticles provide for an array of pathophysiological situations. In this frame, this review addresses general questions concerning the extent to which nanoparticles offer alternatives that improve therapeutic value, while avoid toxicity, by releasing bioactive compounds specifically to target tissues affected by specific chemical and pathophysiological settings. In this regard, to date, the contribution of nanoparticles to protect encapsulated bioactive compounds from degradation as a result of gastrointestinal digestion and cellular metabolism, to enable their release in a controlled manner, enhancing biodistribution of bioactive compounds, and to allow them to target those tissues affected by biological disturbances has been demonstrated.

**Keywords:** phytochemicals; nanoparticles; controlled release; improved functionality; health; smart-formulations

---

## 1. Introduction

In the late 1980s, Dr. Stephen De Felice first coined the term "nutraceuticals" that was defined as "as foods, food ingredients, or dietary supplements with demonstrated specific health or medical benefits, including the prevention and treatment of disease beyond basic nutritional functions" [1]. Nutraceuticals, in general, have been strongly suggested as candidates for the development of chemo-preventive agents, regarding several pathophysiological situations based in the experimental results retrieved from a range of studies [2]. Hence, upon the extensive research developed so far, some of their mechanisms of action that involve competencies to modulate multiple molecular pathways, without eliciting toxic side effects, have been demonstrated [1].

In recent years, and in close connection with the application of the promising chemo-preventive agents, technological advancements have allowed describing nano-biotechnology as a powerful tool to create and enhance the utility of nano-size materials [3], especially concerning the administration of bioactive phytochemicals in the frame of medical treatments [4]. In this respect, diverse types of nanoparticles have been tested for various uses. For instance, nano-silver has been widely used aimed to take advantage of its toxicity against a broad spectrum of microbes responsible, for instance, for superficial skin and nail infections in humans [5], in addition to oral and vulvovaginal corruptions [6]. Based on the recently gathered knowledge, one of the most relevant advantages of implementing this technology is to control the delivery of bioactive compounds, providing new and valuable responses to the current challenges for new "smart-foods", able to promote human health and wellbeing. In this regard, this work reviews critically the recent advances on nanotechnology-related administration of bioactive compounds and its socio-economic implications.

*Societal Impact of Nanobased Technological Innovations for Food and Health*

In recent years, nanotechnology has been noticed as one of the most important fields with clear socio-economic impact in several fields of activity. In this aspect, it has been estimated that the private sector has invested around 150 billion US dollars in the development of nanotechnology and application models during 2015. Regarding this, the number of products that incorporate nanoparticles in their formulations has enhanced, accounting for more than 1800 products containing nanomaterials, produced by 622 companies [7]. This new application alternative has been possible due to the 3.7 billion US dollars investment by USA, upon the National Nanotechnology Initiative, followed by the European Union and Japan, which have invested up to 1.2 billion and 750 million dollars, respectively, per year. If this evolution continues, in 2020, the worldwide economy could be supported by nanotechnologies, with an input of 3 trillion US dollars [8].

Nowadays, and regarding the Science and Technology of Foods, the application by the industry of the knowledge generated regarding nanotechnology is mainly focused on agriculture, foodstuffs transformation, packaging, and storage, and the production of dietary supplements [9]. Indeed, during food processing, the advances achieved so far have focused on nutrient delivery, considering distinct issues that deserve to be depth studied, such as the preservation of the physico-chemical features of foods (taste and color), and their safety, stability, bioavailability, and bioactivity. Regarding this, the use of new nanosystems might result in great interest to achieve additional application for the currently most widely characterized bio-functional compounds. Indeed, nanoencapsulation is a very valuable approach that could allow controlling bioactives' release and their pathophysiological relevance, as well as their preservation after oral ingestion, with relevant applications for human health. As an example of such applications, structural lipids have been explored as carriers of healthy component applied for the inhibition of cholesterol transport from the digestive system to the bloodstream [10]. Thus, resorting to the research results, colloids, emulsions, biopolymers, liposomes, solid–lipid nanoparticles, and nanofibers have demonstrated an indubitable utility as vehicles of functional compounds [11].

Nonetheless, despite the increasing use of nanomaterials and the high number of derived commercial applications, nowadays, there are some concerns regarding the potential risk for human health derived from their use, as well as on the extent in which the inclusion of nanomaterials in foods formulation could compromise foods safety. Indeed, in this concern, one of the main matters of debate nowadays is focused on unravelling the extent to which nanoparticles and the bioactive compounds carried by them are competent to access to tissues naturally protected by biological barriers such as the blood–brain barrier [12,13], as well as the toxicological effects that this transportation could cause that have not been properly addressed so far.

Depending on the nature of nanoparticles, as well as the surrounding environment, the materials currently used can be aggregated, modifying their chemical properties, size, and shape [14]. These changes could result in an altered nanomaterial form and, in this sense, some reports have been focused on the evaluation of nanomaterials toxicity when they form part of a particular matrix [15]. To date, the gap in information existing in these concerns entails a lack of consensus about toxicity and direct/indirect consequences on health. As a result, the legislation governing the use of nanoparticles does not elucidate clearly this situation, being almost limited to the requirement of labeling food products containing nanomaterials. In 2009, the European Food Safety Authority published "The Potential Risks Arising from Nanoscience and Nanotechnologies on Food and Feed Safety" as an official document, while the European Union legislation differs depending if intended use of a specific nanocarrier is as a primary ingredient or constitutes a food additive [16]. However, the soft demand of the industry for incorporating this technology to production chains has prompted to a lack of a regulatory frame for the use of nanotechnologies in foods. In any event, to extend the utilization of this promising technology with industrial purposes, the advance of the desirable features of nanoparticles in diverse food matrixes constitutes an important challenge that deserves to be explored towards an enhanced use of such materials, according to their specific properties, at the time that contribute to preserve safety and bioactivity.

## 2. Biocompatible and Biodegradable Nanoparticles: Looking for the Best Option

### 2.1. Solid Nanoparticles as Attractive Drug Vehicles: Composition and Properties

A variety of nutrients and non-nutrients has been identified, regarding their biological activity once ingested by diet, being responsible for a diversity of biological activities, closely related with the promotion of human health. However, bioactive compounds from nutrients such as lipid derivatives (plant and mammals oxylipins), bioactive peptides, minerals and vitamins, as well as non-nutrients (phytochemicals), can be degraded as a result of the gastrointestinal digestion. Moreover, diverse barriers must be crossed by these compounds, which route towards the bloodstream, and the final distribution to cells and tissues, including their absorption at gastric or intestinal levels. During last decades, the search for effective carriers for bioactive compounds and drugs has been based on the identification of new alternatives for oral administration with health-promoting purposes, avoiding their early inactivation [17,18], because these factors limit the bioavailability of actives or drugs [19]. According to the results retrieved in recent years, nanoencapsulation developed using particles with diameters ranging from 1 to 100 nm enables augmented concentrations of bioactive compounds in cells and tissues, enhancing also shelf-life (reducing the metabolization and excretion of bioactive compounds) through a slowdown delivery [20]. Given these advantages, and aimed to optimize the benefits for health derived from nutraceutical formulations, regarding antioxidant, antiradical, and antitumor abilities, among others, the industry has boosted the research on this technology.

Upon evaluating the profits of implementing nanotechnology in agro-food industries, it has been noticed that nanoparticles have some advantages that deserve to be further explored. For instance, specific high surface zone of nanomaterials has been related to an enhanced number of functional groups, available for chemical reactions, relative to free functional compounds [14]. In addition, the class of relationships between nanocarriers and bioactive nutrients and non-nutrients, and the high area/volume ratios, contribute in some cases to suitable formulations, in which the adhesive properties of the nanocapsules led to prolonged transits of functional compounds through the gastrointestinal tract. This situation lets them reach unaltered specific areas, where their liberation provides enhanced benefits. Consequently, an augmentation of its absorption and bioavailability is achieved [21–23]. Moreover, additional constraints associated to the inclusion of bioactive compounds in the food matrix, such as the effect of light, temperature or oxygen, which compromise their stability and, thus, the bioactivity of the compounds of interest, can be attenuated using nanoparticles [24,25].

Among the nanometric size delivery systems, there are different types of encapsulation, which are classified depending not only on their nature, but also on the production method, properties of the system, the system free energy, the interaction force in the system, etc. [14]. To date, the most suitable nanoparticles with marked potential, regarding distribution of bioactive compounds, are lipid-based delivery systems, which may comprise solid–lipid nanoparticles, liposomes, and micelles, as well as protein- and polysaccharides-based biopolymeric nanoparticles. Specific structural properties of these nanosystems, as well as their low cost and non-toxic nature, make them suitable carriers in food delivery compared to others, such as synthetic polymeric nanoparticles [26]. In addition to nanoparticles, another nanotechnology application for the administration of bioactive compounds are nanoemulsion droplets, formed by dispersion or high-energy and low-energy emulsification methods. Although the main application of nanoemulsions is the preparation where droplets work as nanoreactors, nowadays new specific applications are being envisaged concerning controlled drug delivery and targeting [27].

2.1.1. Lipid Nanoparticles

In the frame of nanotechnology, lipid-based systems are mainly represented by solid–lipid nanoparticles and nanostructured lipid carriers that have been commercially introduced as nanocarriers of functional compounds, in the last years, mainly due to their natural composition and biocompatibility. They can be made of different lipids, including fatty acids, steroids, and waxes to monoglycerides, diglycerides, and triglycerides [28]. A typical percentage ratio fat/aqueous medium of 0.1:30.0 (*w/w*) has been considered for developing solid–lipid nanoparticles [29] that feature spherical morphology and can be visualized, for instance, by transmission electron microscopy [30]. Particle size and stability can be affected by their lipid composition and the surfactant properties.

Within lipid nanoparticles, there are advantages associated to the use of liposomes that constitute nanosize artificial vesicles suitable to be obtained from phospholipids and cholesterol. In this concern, recently, the role of these vesicles as immunological adjuvants and drug carriers has been revealed [30, 31]. Thus, it seems evident that several advantages can be obtained from using liposomes as carriers of bioactive compounds, including their capacity to encapsulate diverse bioactive compounds featured by a range of polarities, which are included into the aqueous core of the phospholipid vesicle or at the bilayer interface, and their structural versatility [32]. Besides, an interesting characteristic of liposomes is that they are obtained from natural lipids, being thus biodegradable, biologically inactive, and without antigenic, pyrogenic, or intrinsic toxicity [33,34]. the compounds within these nanoparticles are preserved from the deleterious activity of external factors, especially concerning enzymes and inhibitors present in the gastrointestinal tract [35]. Based on these benefits, liposomes are increasingly used by the pharmaceutical industry to control the release of several compounds of interest, because of their biological activity (drugs, vaccines, and enzymes) in the prevention of a range of pathological situations [32,36–38]. Nonetheless, liposomes are not absolutely free of drawbacks, as nanocarriers, due to their instability in plasma [39]. In this regard, nowadays, it is required to develop an evasion system to enhance the circulation time of these nanoparticles that would results in an augmented concentration in vascularized tissues [40], especially in cases involving active neoangiogenesis.

Apart from liposomes, micelles are very slim spherical additional lipid molecules of 10 to 100 nm formed in aqueous solutions [41]. These nanoparticles have been revealed as a valuable alternative to improve bioavailability and retention of bioactive phytochemicals, since they provide appropriate protection to these compounds against inactivation reactions by surrounding factors, and constitute an alternative featured by even higher loading capacity and improved stability [41,42]. the release of bioactive compounds from micelles is conditioned by an array of factors, including micelle stability, rate of bioactive compounds diffusion, the partition coefficient, the rate of copolymer biodegradation, the drug concentration within the micelles, the molecular weight, and the physicochemical features of the bioactive phytochemical and its location within the micelles [43,44]. Interestingly, release from micelles can also be enhanced in the targeted area by certain stimuli, such as pH, temperature, ultrasound, and light [45]. Thus, besides providing innovative solution to the solubility and long circulation of bioactive nutrients and non-nutrients, micelles contribute to a more efficient internalization and the proper localization within the separate cell compartments, where the biological activity of the compounds carried is required. This is extremely important because it helps take the highest advantage of the specific mechanism of action from the entrapped bioactive compounds [46], also contributing to higher concentrations at the organic site affected by pathological situations, thus decreasing side effects relatively to systemic administration [47].

Additionally, niosomes are microscopic lamellar structures formed upon non-ionic surfactant of alkyl or dialkyl polyglycerol ether class and cholesterol [48]. These lipid-based nanoparticles are structurally similar to liposomes, constituting an effective alternative to these carriers. Indeed, given their features as non-ionic particles, they seem to be less toxic, while providing an improved therapeutic index for bioactive compounds [49]. Moreover, the niosome vesicles are osmotically active and stable, and, thus, may act as a storage system, providing the controlled delivery of bioactive compounds. Thus, they augment oral bioavailability of compounds with limited absorption, allowing to envisage different therapeutic applications [50–53] and use as diagnostic imaging agents [49].

Solid–lipid nanoparticles share some features with nanoemulsions, such as low cost, good release profile, stability, scalable for industrial production, and food safety [54]. As described for nanoemulsions, when considering solid–lipid nanoparticles, the bioactive compounds are situated in the core of particle; however, higher stability and ability to withhold phytochemicals in the nanosystem complex has been described in relation to emulsions or liposomes [55].

Despite the promising features of solid–lipid nanoparticles, some potential disadvantages have been associated to these nanocarriers [55,56]. Regarding this, the loading capacity is one of the major constraints described so far. This seems to be a consequence of the crystallization or transformation of lipids. Hence, the use of nanostructured lipid carriers could be an alternative to overcome this constraint. Indeed, in nanostructured lipid carriers, although the lipid particles are also solid, the use of specific lipids, such as isopropyl myristate or hydroxyl octacosanyl, avoids crystallization at low temperatures [54].

### 2.1.2. Polysaccharide Nanoparticles

In respect to biocompound delivery systems based on biocompatible materials, it is remarkable that those developed resorting to polysaccharides have been stressed as a valuable alternative to achieve controlled and prolonged release of bioactive phytochemicals [57]. In this aspect, the application of this type of compounds is reinforced by their amphiphilic nature that allows self-assembly in aqueous environments and helps to form specific structures [58,59]. Besides, polysaccharides feature a high affinity to mucosal cell layers in the respiratory and gastrointestinal tracts [60] that turns nanoparticles developed using these compounds into very valuable alternatives to enhance the bioavailability of the bioactive compounds of interest.

Based on the intrinsic charge of polysaccharide materials, they are classified into polyelectrolytes that include cationic, anionic and neutral subtypes, and non-polyelectrolytes [61]. Besides the intrinsic advantages noticed for this type of biocompatible material, it has recently been demonstrated that coat materials, applied on nanoparticles developed using polysaccharides, can interact with specific receptors in cells and tissues. the description of this interaction has allowed envisaging active and site-specific controlled targeting for compounds carried in polysaccharide nanoparticles [60].

Concerning cationic polyelectrolytes, chitosan, which is composed of repeated units of D-glucosamine, has been described as a non-toxic, biodegradable, and bioadhesive material that provides the advantages of control of release of encapsulated agents, complexation with negative charged macromolecules, avoidance of toxic solvents during preparation, and prolonged residence time at the site of absorption due to its mucoadhesive nature [62]. Indeed, the latter is one of the major advantages for the administration of compounds of interests against nasal, oral, ocular, and dermal disorders [63]. Several factors condition the preparation, conformation, and loading of chitosan nanoparticles, affecting their ability to delivery compounds [62]. In addition, chitosan nanoparticles are suitable to be modified by adding specific ligands that contribute to a more rapid and efficient interaction with cells membranes, and thus, to an enhanced delivery of bioactive compounds [64].

In respect to anionic polysaccharide nanoparticles alginate, heparin, pectin, and hyaluronic acid are important to mention [62]. An overview of the interest of these nanoparticles indicates that alginate features biocompatibility, biodegradability, non-antigenicity, and mucoadhesive features [65–67], which make it an interesting candidate for controlled delivery of bioactive compounds [62]. Regarding the mechanism of release, the electrostatic interaction between carboxyl groups and divalent ions mediates the formation of cross-linked gels, with a close relationship with the release of bioactive compounds [62], thus the data reported in the literature indicate these nanoparticles are a valuable alternative delivery system suitable for incorporating positively charged compounds. Interestingly, a combination of alginate and chitosan has been pointed out as an alternative system for the controlled delivery of bioactive compounds that could contribute to extend the circulation time, relatively to alginate and/or chitosan alone [62].

Apart from alginate, hyaluronic acid nanoparticles feature high aqueous solubility and stability, and are non-toxic and non-immunogenic, which have given rise to a reduced number of systems with potential on passive tumor targeting [68], since they exhibit affinity to hyaluronan receptors with a high expression pattern in tumor cells and, therefore, with high potential to carry bioactive compounds, specifically identified concerning anti-tumor activity, that could release in the precise site, contributing to increase their bioactivity.

Finally, in respect to neutral compounds, represented by dextran, pullulan, and pectin [62], the presence of hydroxyl groups modulates the incorporation of bioactive molecules in the base skeleton. Within this class of nanoparticles, dextran, for instance, has been used to design delivery systems able to escape the reticuloendothelial system, thus enhancing the circulation time of the bioactive compound of interest [62].

### 2.1.3. Protein Nanoparticles

To date, diverse arguments have been proposed supporting the use of protein nanoparticles: (1) simple manufacturing by just heating protein solutions; (2) absence of requirement of emulsification performance; and (3) compatibility with the high pressure emulsification process. Moreover, protein nanoparticles feature high freeze–thaw stability, since the particles at the interface provide good stearic stabilization [69]. To date, different types of nanoparticles have been described, which are synthesized resorting to diverse classes of molecules. Actually, protein nanoparticles can be obtained with a plethora of lab approaches, each of them with its advantages and constraints. However, when compared with additional compounds suggested as nanoparticles for carrying bioactive compounds, protein-based nanoparticles have various advantages, namely the abundance of proteins in nature, their suitability to be transformed, and the absence of deleterious effects concerning the biological systems in which they are applied.

the first classification of protein-based nanoparticles refers to their origin (animal or plant proteins). the diverse types are associated to benefits and constraints, mainly regarding toxicity and/or infections associated to their application (animal proteins), while plant proteins provide the advantage of their hydrophobic characteristics because, which is associated with a lack of toxicity and a lower economic cost [70]. Besides, the surface features of protein-based nanoparticles are susceptible to be modified due to the occurrence of functional groups, which is of crucial relevance for obtained desired biodistribution, biocompatibility, molecules carrying capacity, and stability [70]. In addition, the surface amino groups, characteristics of proteins, allow the addition of hydrophilic polymers, such as polyethylene glycol that augment the circulation time [71]. the loading efficiency of protein-based nanoparticles is closely related to the isoelectric point. When assessing the compounds loading to the protein nanoparticles, it is noticed that the total amount of bioactive compounds is determined by monitoring (ultraviolet (UV)-spectrophotometry, fluorescence spectrophotometry or high performance liquid chromatography (HPLC)) the non-entrapped compounds present in the supernatants. This constitutes critical information for setting up the correct administration of the bioactive compounds. Additional features that should be considered regarding protein-based nanoparticles are the delivery of the compounds carried and the biodegradation of the nanoparticles, being both interdependent reactions [69].

### 2.1.4. Nanoemulsions

As mentioned before, the application of nanoemulsion droplets, to the controlled delivery of bioactive compounds, is a recent approach in the field of Food Science and Technology, thus still has much further to go [27]. To date, nanoemulsions have been mainly applied to the design of new functional foods, since this technology allows regulating the release of bioactive nutrients and non-nutrients featured by poor water solubility [72,73]. For the preparation of nanoemulsions, aqueous solutions of lipid droplets (size approximately 100 nm) are prepared using different methodologies. the most common technology applied is high pressure homogenization [74,75] that avoids particle aggregation and compounds separation by gravitation in comparison with normal emulsions [76]. In fact, in nanoemulsions, the functional compounds are protected from reacting with the food matrix, which helps to preserve organoleptic properties, while enhancing the bioavailability of the bioactive compounds of interest, because of favoring passive transport through biological membranes [77,78]. Afterwards, during digestion, it has been reported that nanoencapsulation increases solubility of bioactive nutrients and non-nutrients, as well as their presence suiting aborbtion at the gastric and intestinal levels. Moreover, the application of these compounds using nanoemulsions provides enhanced protection, reducing their metabolism and lowering the activity of efflux transporters [79,80].

Currently, there is an extensive literature on the encapsulation of functional compounds, such as curcumin and resveratrol [73,81–83], using nanoemulsions to protect these compounds from digestion. Indeed, despite the evidence retrieved through the last decades from the array of in vivo and in vitro characterizations, developed on the benefits for human health of bioactive phytochemicals, their current use is still limited due to poor bioavailability [84]. Regarding this constraint, an adequate formulation of nanoemulsifier has been demonstrated as a valuable alternative to improve the protection of these compounds from chemical and enzymatic aggressions in the intestinal lumen, and thus provide an enhanced absorption and bioavailability. In this sense, combinations of lipophilic and hydrophilic emulsifiers, specifically nanoemulsions composed by lecithin, are competent to entrap resveratrol in the nanoparticle core by hydrophobic forces, enhancing not only the shelf-life of the bioactive molecules, but also their bioavailability by increasing the intestinal absorption and cellular uptake [85].

In addition to the specificity of the nanoemulsions for the diverse bioactive compounds, their stability is also a critical issue that has been deeply evaluated in recent years. Hence, the stability during storage is conditioned by an array of factors, namely pH, temperature, and ion strength, but also on some intrinsic properties of the formulation as the emulsifier concentration or the use of additives that should be adjusted carefully to guarantee the appropriate functionality of nanoemulsions [75,86].

Another factor that has to be considered to achieve the best outcomes from nanoemulsions, as carriers of bioactive compounds, is the nanoparticles size. the right nanosize of emulsions could be reached using co-adjuvants, since droplets prepared from low viscosity oils have been shown to be able to reach the nanoscale. An example of this is the addition of butanol to sodium caseinate that presents a nanoscale after the addition of the alcohol [87]. Polyethylene glycol and ethylene glycol have also been used to achieve the desired nanomagnitude.

From the technological point of view, in the beverage industry, an additional quality of nanoemulsions, i.e. the high optical clarity, provides advantages during manufacturing that have advised on the challenge of prioritizing this technology in the design of new added-value products [88]. However, it is required to state that each protein has the capacity to encapsulate specific compounds according to their hydrophobicity or hydrophilic features.

### 2.2. Nanotechnology for Medical and Nutrition Research

As mentioned above, dietary supplements are useful for the prevention or reduction of different pathophysiological situations and diseases, supported by their antioxidant, anti-inflammatory, and antitumoral properties, among others [89]. Encouraged by the extensive bibliography on these biological activities, many supplements have included phytochemicals in their composition. Given diverse troubles reported regarding bioavailability (that is closely linked to their biological activity in vivo), to take advantage of their biological potential, research efforts have been addressed to explore the diverse encapsulation modes that would enhance bioavailability. Although this approach has allowed developing new nanoencapsulated products addressed to be administrated via ocular, transdermal, and intravenous [90,91], their application in the design of functional foods, because of the specific chemical and enzymatic conditions in the gastro-intestinal compartments, with a direct impact on their release from the food matrix after oral ingestion, integrity, and functionality, needs to be further explored [92].

As an example of these applications, nanoemulsions have been successfully used for curcumin (diferuloylmethane) and dibenzoylmethane (a structural analog of curcumin) encapsulation [93]. These compounds have demonstrated valuable therapeutic potential, while exerting reduced toxicity, in phase I human clinical trials [94]. Upon the studies developed to improve their bioavailability and bioactivity, it has been found that using a mix of triacylglycerol (as oil) and Tween-20 (as emulsifier) could be enhanced the cellular uptake of curcumin and, consequently, an amended anti-inflammatory activity [93,95]. Hence, although other encapsulation alternatives have been considered for curcumin, such as hydrophobic starch [96], albumin [97], β-lactoglobulin [98] or chitosan [99], the poor water-solubility of these compounds limited their application, remaining nanoemulsions as the most suitable and efficient encapsulation alternative.

Other polyphenols of interest have been nanoencapsulated in emulsions for experimentation, such as tannin, stilbenes, and flavonoids [100]. This reinforces the idea of nanoemulsions as a valuable choice for these types of bioactive compounds. However, the emulsion composition has been reported as a decisive factor for the final bioactivity. Thus, epigallocatechin gallate, a hydrophilic flavonol, featured by similar and low bioactivity as antioxidant, in mice tissues, when formulated in two esterified emulsions or in an aqueous phase, while the selection of a third emulsion composition results in an increased antioxidant activity in mice plasma [101].

Essential oils are other important plant derived bioactive compounds with interesting benefits, regarding the prevention of chronic pathologies or infectious diseases. Indeed, based on this evidence, these compounds have been applied as analgesic, sedative, anti-inflammatory, spasmolytic, and local anesthetic treatments [102]. Thus, although a great variability in micro encapsulation methods has been described for these compounds, such as polysaccharides, spray-dried powders or Ca-alginate [103,104], these encapsulation alternatives provide protection to essential oils but seem to be no competent to enhance their antimicrobial activity. However, the nanoscale promoted a slower delivery and higher cell permeability, especially in the skin layers, increasing the activity of essential oils [105].

According to the information referred above, lipid-based nanocarriers are noticed as valuable alternatives to entrap essential oils in the core of the nanostructure and thus, reach different type of cells. In this regard, these nanoparticles have been applied in microbial infections due to their higher cell penetrability [106]; however, in some cases, antimicrobial capacity was reduced when compares with microencapsulation forms [24].

Efficient nanoencapsulation of other compounds, such as ferulic acid and tocopherol in solid–lipid nanoparticles, has been also reported recently [107]. Thus, in nanoemulsions, bioactive lipids were protected against autoxidation [108,109] and carotenoids [110], increasing their bioaccesibility.

Another functional compounds used in the formulation of health foods are probiotics, microorganisms, that improve intestinal microflora. Regarding probiotics, targeting specific regions in gastrointestinal tract after nanoencapsulation has been achieved [111]. Nanoemulsions were used to protect lactic acid bacteria [112] and, thus, to fine-tune the microbiota in the diverse intestinal conditions.

the micelle delivery system, for instance, is widely accepted as valuable carriers of antitumor drugs, and has been evaluated in the frame of clinical trials [43]. In this regard, hydrophobic drugs, for instance, can only be administered intravenously after their conditioning with solubilizing adjuvants, which has been associated with the appearance of toxic symptoms [113]. However, the addition of these drugs into nanocarrier-like micelles allows the replacement of the toxic adjuvants and contributes to minimize toxic effects [114]. Furthermore, as an additional advantage, the unimolecular micelles exhibit a maintained distribution jointly with a controlled pH-sensitive drug release. Likewise, it has been demonstrated that rapamycin could be loaded efficiently in mixed micelles up to a concentration of 1.8 mg/mL by using a hot-shock protocol. Upon these characterization, the release kinetic of obtained for rapamycin indicates that this micellar system could be triggered by varied pH environments under physiological conditions [115]. Matsumura et al. demonstrated that a paclitaxel micellar formulation consisting of polyethylene glycol and modified polyaspartate as hydrophobic block features cytotoxicity in a range of human tumor cell lines compared with the administration of the bioactive compound alone [116].

## 3. Phytochemicals Loaded Nanoparticles: Diving into Nanosized Drug Delivery Systems

### 3.1. Bioavailability Advantages of Nanoencapsulated Phytochemicals

Phytochemicals feature a diversity of chemical structures, represented by phenolic acids, indoles, alkaloids, isothiocyanates, phytosterols, saponins, and phytoprostanes/furanes [117,118]. These compounds have received growing attention boosted by their benefits for human health [119,120], upon a plethora of mechanisms including free radicals scavenging, inhibition of the assembly of microtubule and microfilament, metal chelation, and/or protease inhibition, among others [121]. Hence, the assessment of these compounds, from their in vivo metabolization point of view, has revealed that, after oral intake, phytochemicals are recognized and processed as xenobiotics and degraded as a consequence of the different digestion chemical, enzymatic, and microbial phases in mouth, stomach, and small and large intestine; then absorbed, distributed, and metabolized in cells and tissues; and finally excreted via renal, biliary or pulmonary [122]. Meanwhile, the maximum bioaccessibility and bioavailability of phytochemicals give rise to their actual capacity to exert biological effects in vivo (bioefficacy). the bioactivity of the bioactive nutrients and non-nutrients is demonstrated by short-term changes in the expression of biomarkers of liver function, plasma lipid profiles, blood pressure, plasma glucose, and plasma antioxidant activity [123]. In this sense, bioefficacy is influenced by the chemical nature of these compounds and their metabolic derivatives.

the innumerous articles published to date on the bioavailability of phytochemicals thereafter their oral intake have demonstrated that these compounds or their metabolites may fall in the nano/picomolar range in cells and tissues, which may result in insufficient dose to achieve the biological efficacy, demonstrated upon in vitro characterizations [124].

the dispersion and absorption, of these compounds, is closely dependent on their polarity. Once absorbed, small intestinal enterocytes are responsible for the cellular uptake, efflux pumping, and phase I and II metabolism, while the results of these processes are responsible for the amount of bioaccessible phytochemicals [122]. Moreover, transformations occurring in the frame of phase I and II reactions in hepatocytes give rise to active formats from inactive precursors that are the actual responsible of their biological value [125]. In addition, unabsorbed phytochemicals reach the large intestine, where are further metabolized by the local microbiota. This additional modification also affects their bioefficacy, providing new bioactive compounds that contribute to the final biological interest of a given plant matrix [122].

To prevent the deleterious effect of metabolization on the occurrence of truly bioactive compounds in cells and tissues, at operative concentrations, it has been pointed out that edible nanoencapsulation vehicles could be a valuable alternative. Regarding this, nanoencapsulation technology has been implemented, based on the information retrieved from an array of in vitro and in vivo characterizations, set up aimed at assessing the capability of different tissues and cell types to uptake the bioactive compounds at operative concentrations, according to an array of pathophysiological situations.

Diverse safe ingredients including lipids, polysaccharides, proteins, and biodegradable polymers are currently applied towards the development of edible nanoparticles featured by well-characterized size, surface properties, matrix materials, and compartment structure. These nanoparticles are developed foreseen to obtain improved strategies to enhance the release of the compounds of interest in target cells and tissues, as well as an increased bioefficiency [126,127]. the diverse types of nanoparticles suitable to be used as carriers of bioactive phytochemicals, in the frame of oral administrations, display a variety of structural features, such as is exemplified by nanoliposome, micelle, nanoemulsion, solid–lipid nanoparticles, polymeric nanoparticles (nanospheres and nanocapsules), protein–polysaccharide complex coacervation, cyclodextrin inclusion, and polymeric nanogel. Indeed, these alternative nanoparticles provide a range of features, including the process of loading phytochemical compounds, encapsulation efficiency, stability during gastrointestinal digestion, the releasing mechanisms and, therefore, their capacity to enhance the bioefficacy of the carried phytochemicals [122].

This situation has been recently illustrated by the evaluation of neuroblastoma cells (SH-SY5Y cell line), on their ability to uptake nanoparticles that provide promising results. In fact, the results retrieved from this work revealed that nanoparticles contribute to reach efficient concentrations of the bioactive compounds in the nervous system, especially when applying anionic charged nanoparticles, that exert a higher capacity to successfully go-through the blood–brain barrier, in comparison with cationic nanoparticles [128]. In addition, when using this type of nanoparticles, given the repulsion among the high negatively charged nanoparticles, it is possible to achieve extra stability for the bioactive compounds in cells and tissues, in vivo [129].

A careful revision of the literature gives a close relationship between the capacity of nanoparticles to enhance phytochemicals' bioefficacy with particle size (the most relevant feature), surface properties, matrix materials, and compartment structure [130]. Actually, these features affect bioefficacy, influencing dispersion and gastrointestinal stability, as well as the release rate and the delivery site, the efficacy of the transportation through the endothelial cell layer, the systemic spread, and their capacity to control the impact of the microbiota metabolism [122]. Besides, when targeting the improvement of cellular uptake, it is essential to notice that particle size, jointly with the use of polymeric nanoparticles, solid–lipid nanoparticles, or nanoemulsions, contribute to the stability by affecting to the repulsion facts and/or interfacial-tension-decreasing compounds [131,132], which contributes to augment cellular uptake.

Based on these advantages, loading phytochemicals within such edible nanoparticles enhances the dispersion of initially water-insoluble phytochemicals in aqueous media [133], while the physical barrier provided by nanoparticles protects bioactive compounds against oxidation under acidic and alkaline degradative conditions, in stomach and small intestine, respectively [134,135].

In the literature the bioactivity of phytochemicals in their native state has been frequently reported, while, in vivo, as mentioned before, the delivery of bioactive nutrients and non-nutrients are critically affected by their solubility, which affects stability and penetration under gastrointestinal conditions. To overwhelm this conflict is a basic objective of nanotechnology in the field of Science and Technology of Foods, specifically regarding the development of functional foods and nutraceuticals. Hence, encapsulation of phytochemicals constitutes valuable alternatives to release bioactive compounds in specific tissues, affected by pathophysiological situations. In this sense, carrying bioactive compounds on nanoparticle limits their degradation under the chemical and enzymatic conditions of the gastrointestinal tract, as well as their general metabolism. In this regard, the physico-chemical features of the materials used to develop nanoparticles, as well as the digestibility of nanoparticles in the gastric or intestinal sections, are key issues that should be addressed, firstly, when designing a nanoparticles-based strategy to the oral administration of phytochemicals [122].

Some examples illustrate this situation: while starch-based nanoparticles are digested at oral level by the activity of $\alpha$-amylase, additional polysaccharide nanoparticles are degraded in the small intestine and protein–polysaccharide nanoparticles can be due to variations of pH and salt concentrations, such as those occurring in the gastrointestinal lumen during the separate stages of digestion [135]. Lipid nanovesicles release phytochemicals in the small intestine simultaneously to the digestion of triglycerides [122].

In addition to the nature of the nanoparticle, the loading method has been revealed to be critical for the final release of phytochemicals and, thus, for their bioavailability and bioactivity. Regarding this, the drug-loading method strongly influences the loading efficiency, and it has been described that nanoencapsulated compounds release to a lower extent relative to nanosphere-based delivery, even if both formulations have similar efficiency [90].

Recently, a new type of compounds has emerged, plant oxylipins, such as phytoprostanes and phytofurans, which are formed by non-enzymatic oxidation of α-linolenic acid. Bioavailability and biological activity of such compounds have been suggested based on their structural analogy with human eicosanoids [118,136]. These lipid phytochemicals are compounds of special features as they migrate into fatty acids-based micelles, which contribute to their solubility and subsequent cellular transportation [122].

Apart from the control of phytochemicals release, nanoparticles also influence the transport of bioactive compounds through enterocytes (by transcellular endocytosis for particles between 20 and 1000 nm), by modulating residence time, transportation efficiency, and pathways, as well as the metabolic reactions undergone by such compounds within endothelial cells. Hence, for instance, chitosan derivatives have been related to mucoadhesive features and their capacity to open tight junctions, improving paracellular transport [137–139]. Moreover, the nanoparticle's charge also influences the formation of hydrogen bonds with the mucosal surface, contributing to a momentary retention [140]. Another feature of nanoparticles that influences their role as carriers of bioactive compounds is the incorporation of cell-penetrating ligands that could contribute to enhance transmembrane transport efficiency [122].

Once absorbed from the intestinal lumen, nanoparticles loaded with bioactive phytochemicals move through endothelial layer to migrate to tissues into bloodstream. At this stage, nanoparticles have already been included in endosomes and undergone oxidation, reduction, and hydrolysis reactions, among others, as well as conjugations by diverse metabolization routes upon phase II metabolism. However, unlike metabolism of free phytochemicals, when carried on nanoparticles, the bioactive compounds are protected from such reactions, by modulating the exposition to metabolizing enzymes in enterocytes [141].

Highly lipophilic phytochemicals featured by specific features, regarding the logarithm of the oil−water partition coefficient and a long-chain triglyceride solubility (e.g., phytoprostanes and phytofurans), may go through enterocytes and form chylomicrons with enterocyte lipoproteins [142]. These reach the bloodstream via mesenteric lymph and thoracic ducts, avoiding hepatic first-pass metabolism. This route of absorption has been imitated using nanoparticles incorporating a lipid phase (for instance, emulsions, liposomes, and solid–lipid nanoparticles), which enhance the participation of the lymphatic absorption at the time that augment the rate of transcellular absorption [143]. Given the interest of integrating lipid phase in nanoparticles as modulators of the absorption rate, the extent to which the phospholipid composition and concentration of emulsifiers are competent to fine-tune biodistribution by modifying the lipophylicity and plasma-binding properties of phytochemicals has been investigated [144].

Additional approaches that have been explored, as valuable modes to control release and bioavailability of bioactive nutrients and non-nutrients administered, using nanoparticles as carriers are represented by the use of magnetic nanoparticles, the alteration of the surface chemical properties, and the inclusion of ligands facilitating the cellular uptake [122]. However, for some types of compounds, such as stilbenes, flavonols, and anthocyanins, which are metabolized by the intestinal microbiota towards even more powerful bioactive compounds, could be interesting for avoiding their digestion and absorption at gastric and small intestine levels. With this objective, the best option lies in the use of nanoparticles manufactured with materials or multilayer structures, resistant to gastric and small intestine conditions in terms of pH, osmotic conditions, and enzymatic activities. In addition, to achieve this objective, it is required to design release mechanisms that allow the delivery of bioactive phytochemicals under the osmotic and pH conditions of the large intestine, in which the microbiota responsible for its transformation is present [122]. On the other hand, once this objective is achieved, the challenge of reaching highly bioavailable compounds, derived from the microbiota metabolism, constitutes an additional constraint, due to the low colonic absorption at the conditions of luminal fluid volume, viscosity, and neutral pH of colon [145,146].

Once absorbed and circulating in blood stream, diverse strategies have been suggested and explored to control the length of circulation, such as using hydrophilic coating. Polyethylene glycol has been demonstrated as reliable to enhance circulation time, by acting as a steric barrier and thus protecting nanoparticles from opsonization [147].

One of the most widely used material for the development of nanoparticles is poly-(lactic-coglycolic acid) that represents a successfully used biodegradable polymer due to the production of the metabolite monomers lactic acid and glycolic acid during their physiological hydrolysis, that are easily processed via the Krebs cycle. Due to this rapid metabolization, low systemic toxicity has been described associated to the application of this biomaterial for the control of the bioactives compounds release [148]. However, it is crucial to develop surface modifications using nontoxic and blood compatible material, that allow their uptake by macrophages and the augment of the length of blood circulation, which contribute to the sustained delivery of the bioactive compounds carried in this nanosystem [149,150]. In this connection, poly(ethylene-glycol) is used as hydrophilic nontoxic segment in combination with hydrophobic biodegradable aliphatic polyesters, because this provides the capacity to resist against opsonizing and the ulterior phagocytosis, contributing to enlarged shelf life in the bloodstream and tissues [151–153].

### 3.2. Nanotechnology for Bioactives Delivery

In recent years, nanotechnology has emerged as a valuable alternative to control drugs delivery and tissue engineering [31], because of the chance provided by this to design and develop new strategies that could enhance the efficacy of natural bioactive molecules, currently used in a very limited extent, as therapeutic compounds for the treatment of different pathophysiological situations [32]. Indeed, nanotechnology is an alternative to traditional formulation, which could contribute to improve general bioavailability and site-specific release, simultaneously to toxicity reduction [33–35]. According to these aims, several epidemiological studies have been developed, during the last years, aiming to take higher advantages of the influence of dietary habits on the incidence of an array of pathophysiological situations, that have focused the assessment of several plant foods. From such work, multiple compounds that display potential health benefits in vitro have been identified [154,155]. However, when evaluating the activity of those compounds upon in vivo studies, many of them were not competent to translate the activities previously demonstrated in vitro, appearing as unstable in the intestinal environments and, thus, exhibiting poor bioavailability. Thus, to achieve the required bioavailability rates to take advantage of their biological activities, as a first approach, higher doses were tested, which showed efficacy but resulted in systemic toxicity [95].

the major outcomes from these studies have allowed to notice that several factors are involved in bioavailability, viz. chemical structure, solubility, stability against gastric and colonic pH, metabolism by gut microflora, absorption across the intestinal wall, active efflux mechanism, and first-pass metabolic effects [147]. However, when reviewing the data reported on the bioavailability of phytochemicals, it is revealed that many of them are only partially absorbed/metabolized by the intestinal epithelium, a cellular type responsible for systemic occurrence by their uptake using efflux transporters (mainly P-glycoprotein) [156]. the efflux transporter proteins mediate the active extrusion of compounds back into the intestinal lumen [157] that, in turn, limit the bioavailability of the bioactive phytochemicals of interest to nanomolar concentrations in blood, even if the ions have been detected the intestinal section, where they are absorbed, in high concentrations [158]. Indeed, this situation demonstrates that it is difficult to assess the bioavailability of chemical agents based solely on their physicochemical properties.

This is not just a general concept related to the biological activity of a given phytochemical compound, the bioavailability of the compound itself, at the target site, is a major issue that has not be appropriately addressed in the last years [159]. In this regard, understanding the concerns enclosed to the bioavailability of an individual compound is essential to draw strategies that might help to overwhelm these limitations. In this sense, the emergence of new technologies that allow the advance towards efficient and safe administration of newly identified bioactive compounds, has increased the interest in new "smart" delivery systems that contributes to improve the pharmacological properties of the administrated compounds [147]. Hence, a promising approach to avoid low bioavailability and systemic toxicity, associated in some extent to the use of xenoparticles as carriers of the bioactive compounds of interest, is the application of nanoparticles manufactured using materials of demonstrated safety, such as polymeric nanoparticles, liposomes, dendrimers, and micelles [160,161]. Actually, the use of these types of carriers is associated to innumerous advantages relatively to the traditional systemic administration, since this approach constitutes a powerful tool to modulate the pharmacokinetics and to improve delivery of bioactive agents to target sites [147].

Indeed, nanoparticles may enhance the oral bioavailability of poorly soluble compounds, as well as the tissue uptake after parenteral administration, supplying enhanced adherence capacity to the capillary wall during the initial diffusion phases (tethering, rolling, adhesion, and transmigration). Moreover, the selection of the most appropriate options between the currently developed could also allow augmenting the delivery across membranes and biological barriers [162]. In this regard, the size limitation to go through diverse biological barriers is closely related to target location and the chemical features of the tissue [163].

In recent years, stimuli-responsive polymer-based nanocarriers have focused the attention of researchers, working on the evaluation of bioactive phytochemicals in vivo. Regarding this, such nanoparticles are exposed to modification of their physical and chemical properties when exposed to the specific conditions (pH, temperature, light, magnetic field or glucose levels) in the intestinal tract or the diverse tissues after distribution [163,164]. This dependency could be of significant interest, as they provide the chance to set up valuable relationships between the release of bioactive compounds and specific niches or pathological state [132]. This situation is exemplified by paclitaxel delivery once loaded on pH-responsive nanoparticles, which has evidenced its utility upon in vitro and in vivo antitumoral determinations [165,166].

Anthocyanins have been highlighted, for several years, as model phytochemical to assess the potential of competence of polymeric nanoparticles to encapsulate hydrophilic compounds. the final objective of these studies is to establish the contribution of these approaches to the bioavailability and controlled release and release kinetics in vitro. In the case of anthocyanin, these features have been evaluated on polyethylene glycol moieties. Hence, in these works, anthocyanins were encapsulated with 60% efficiency in biodegradable nanoparticle formulation based on polylactide-*co*-glycolide, which were stabilizer resorting to the use of polyethylene glycol. These nanoparticles evidenced a biphasic release profile, in vitro. In addition, all the polyethilineglicosylated nanoparticles present a similar delivery pattern, characterized by an initial abrupt release followed by a continued supply [167]. Hence, this work described that bioactive compounds may be adsorbed on the surface of nanoparticles as a burst release that is not maintained in time, while the sustained release is due to the liberation of compounds encapsulated in the core domains of the nanoparticle [168].

### 3.3. Nanoparticles towards Targeted Bioactivity

A further step of what has been discussed above is the exploration of the advantages provided by nanotechnology concerning the release bioactive compounds carried in nanoparticles in specific tissues or cell types affected by pathophysiological events that needs to be treated. In the last decades, it has been boosted the application of bioactive phytochemicals as promising compounds capable to prevent health disturbance by modulating molecular pathways in cells. These compounds feature strong radical scavenging, anti-inflammatory, and neuroprotective activities, among others. Nonetheless, most bioactive nutrients and non-nutrients are unstable and rapidly transformed into other compounds with different biological attributions, which jointly with their fast degradation and excretion, limits the actual benefits that could be retrieved from such healthy compounds [169]. For instance, during digestion, the hydroxyl radicals of phenolics (for instance, anthocyanins) are oxidized into quinones, reducing the biological power of these molecules [170]. Traditionally, these constraints have been overwhelmed by therapeutic alternatives outlined for several diseases, which include invasive in situ administration of bioactive drugs, as a way to guarantee high tissue concentrations of the functional molecules of interest. Nevertheless, these administration alternatives are characterized by reduced patient compliance, because of the distress associated with frequent administrations, generally related to an array of side effects [171,172]. Alternatively, substitutive application forms of bioactive phytochemicals have been associated to augmented costs, particularly when repeated administrations are required to complete the treatment for a specific pathophysiological situation [173].

Apart from the side-impacts of the traditional administration of bioactive compounds, a loss of bioactivity during storage, due to the xenobiotics metabolism in mammals, has recently been reported, which could prevent them from reaching the target tissues and cells in which their activity is desired [174–176].

To date, diverse works have been developed aimed to identify alternatives that allow to overcome these problems; for instance, by using high dose or multiple treatments of bioactive compounds. However, this has been associated with dangerous side effects related to overdoses due to nonspecific toxicity of such compounds [174]. As a result, it has been identified an urgency to develop biocompatible and biodegradable systems for packaging phytochemicals, based on carrier systems constituted, for example, by oil-in-water emulsions and liposomes [177]. These alternatives have been demonstrated useful to provide sustained release of therapeutics in target tissues and, in addition, to reduce biological disturbances in cells and tissues, no compromising the stability and functionality of macromolecules [178]. Hence, this provides stability that in turn is responsible for a higher bioavailability and long term circulation [178], due to the protection supplied against outer stresses in vivo [179]. the research on controlled delivery options in respect to tissue, concentration, and pathophysiological modifications, has given rise to the description of biodegradable polymers and formulations specific for diverse types of bioactive compounds, with interesting biological activities, which have focused the preparation of phytochemicals-encapsulated nanoparticles [180]. Based on these applications, this technology has been noticed as highly useful for controlled delivery of bioactive compounds in target sites.

Moreover, the structural features of the nanoencapsulations pointed out so far, have a direct impact on the release profile [90,181]. This is specifically referred to as core−shell nanoparticles, double-emulsions, gelled networks, multiple coating systems, and "prodrug delivery systems". Actually, when these systems are conjugated with azo- or glucuronic acid-including polymers, it is possible to control the bioactive compounds carried by enzymatic hydrolysis, which can contribute to retard the speed of phytochemicals release, as well as a noticeable change of their physical dispersion state [123].

This is of special relevance regarding compounds that develop their biological functions in tissues with high protection against xenobiotics by biological barriers, such as the blood–brain barrier. In this frame, anthocyanins contribute to improve brain functionality and to reduce oxidative stress associated to normal cells metabolism, resorting to their radical scavenging, anti-inflammatory, and anti-neurodegenerative capacity [168,170], thus preventing memory losses in estrogen-deficient rats [182]. the polymer-based nanoparticles used for characterization in this model were featured by a biphasic release profile, providing enhanced neuroprotective power of anthocyanins against Alzheimer's dementia. Besides, the efficiency of the anthocyanins administered associated to nanoparticles has been further demonstrated by monitoring the capacity to attenuate the expression of clinical (amyloid precursor protein and beta-site amyloid precursor protein cleaving enzyme-1), inflammatory (p-nuclear factor kappa-B, Tumor Necrosis Factor-alpha, and nitric oxide synthase), and apoptotic (B-cell lymphoma-2 (BCL2), BCL2 associated X (Bax), and caspase-3 protein) markers [170]. Using this model, it was evidenced that anthocyanins loaded nanoparticles reduced significantly the level of protein markers and were also more efficient in modulating the P38/JNK pathway, according to reduced expression of various inflammatory markers, cytotoxic compounds, and proinflammatory cytokines [168]. In respect to oxidative stress, non-conjugated anthocyanins or molecules associated to nanoparticles significantly upregulate endogenous antioxidant genes, such as nuclear respiratory factor-2 and HO-1 proteins, which has been related with the prevention of oxidative stress and consequently, with an attenuation of the clinical symptoms of the Alzheimer's dementia [183].

An additional functionality demonstrated for anthocyanins loaded nanoparticles is related with their capacity to revert the augmented and decreased expression of pro-apoptotic (caspase-3 and Bax) and anti-apoptotic proteins (BCL2), respectively, thus reducing DNA damage in a higher extent than native non-conjugated anthocyanin. All these findings together prompted Amin et al. [168] to suggest a neuroprotection activity of anthocyanins loaded nanoparticles that surpassed significantly the biological potential, enclosed to free anthocyanin. This enhanced activity, when combined with polysaccharides, seems to be related to an increased stability and prolonged degradation time [170].

**4. In Situ Bioactive Compounds Delivery Control: Drawbacks and Breakthrough Advantages**

According to the constraints outlined before, in the last decades, growing attention and resources have been devoted to the development of carrier systems that allow local delivery of therapeutic agents, for instance, by using organic materials. Regarding this, control delivery has been identified as a key stage of bioactive compounds administration, foreseen as a way to set up the timing, tissue/cell, and pathophysiological conditions, under which therapeutic agents are released. Indeed, achieving this objective would allow reaching higher local concentration near to the operative levels demonstrated in vitro, while reducing the overall administered dose (and consequently systemic toxicity associated). According to this objective, as a result of the research efforts developed in the last years, a diversity of internal and external factors can control the specific release of bioactive phytochemicals, according to a range of factors, including pH, the activity of local, temperature, ultrasound, magnetic field, and/or light incidence [184].

Moreover, recently, the nanoencapsulation of compounds identified as potential therapeutic agents has raised growing interest, due to the augment of the range of biomaterials with valuable applications in this field [163,185]. Thus, polymeric nanocarriers can increase the bioavailability, improve solubility, and prolong the shelf-life of potential compounds which, to date, have been difficult to deliver in a controlled way. For instance, oil-cored nanocapsules improve originally the administration of bioactive compounds, allowing targeted delivery and controlled, long-term, release that contribute to decrease the dosage and frequency of administration giving rise to an increased patient compliance [184]. Due to this, the layer-by-layer self-assembly of pH-sensitive building blocks has been explored as a promising approach to obtain biomaterials with customized properties [186] and, thus, with interesting applications as stimuli-responsive nanocarriers for controlled release of phytochemical compounds, which could also provide transport capacity through diverse biological barriers [187]. Moreover, the use of appropriate pH-dependent biocompatible polyelectrolyte for nanocapsule shell formation can provide the tools required to design non-toxic nanocarriers, with shell permeability [184].

Loading bioactive phytochemicals in nanoparticles turns bioactive compounds into more effective and contributes to achieve site-specific delivery. Once into the cells, some nanoparticles release the bioactive phytochemicals slowly, contributing to sustained therapeutic effects, which constitutes an additional advantage [188]. This capacity was demonstrated recently by using rhodamine-loaded polyethylene glycol-nanoparticles. Upon this characterization, after the application of rhodamine labeled nanoparticles to SH-SY5Y neuroblastoma cells, red fluorescence is visualized in the cytoplasm, suggesting that nanoparticles are internalized via endocytosis [189], saving the phospholipidic barrier that the cell membrane represent for hydrophilic compounds. An additional demonstration of these advantages was obtained on prostate cancer DU145 cells with similar positive results [190].

In addition, in relation with the administration of bioactive compounds, pursuing the development of activity at neural tissues and in the frame of deficits or overexpression of neurotrophins, responsible for neurodegenerative diseases and psychiatric disorders, to date, the research has demonstrated a delivery system competent to control of neurotrophin dosage in the brain. In fact, the major outcomes obtained in this issue have suggested that carrying bioactive compounds in nanoparticles might favor targeted delivery in specific brain areas, minimizing biodistribution to the systemic circulation and, consequently, toxic side-effects. Indeed, this approach provides valuable benefits concerning neuroregeneration [191].

In addition to oral administration, polymeric microspheres, scaffolds, and conduits have been used as sustained-release systems for neurotrophic proteins [192,193] that even provide enhanced neuroregeneeration, relatively to the implantation of polymeric scaffolds resorting to invasive surgery techniques into the central nervous system [191]. Indeed, this situation is exemplified by a slow release achieved of glial cell line-derived neurotrophic factor, by-using micro-reservoirs created by biodegradable poly-(lactic-*co*-glycolic acid) microspheres or by Poly(lactic-*co*-glycolic acid) microparticles [194,195].

Although the actual significance and advantages of the abrupt release of bioactive compounds, when using controlled delivery systems, has not been entirely ignored over the last years, to date, no plenary successful theories have explained completely the phenomenon. In addition, it has to be considered that the negative effects, associated to burst release, are pharmacologically dangerous and economically inefficient [196]. To overwhelm this inconvenience, new nanoformulations composed of thermo-sensitive gelling copolymer has been successfully formulated and characterized. the negligible amount of bioactive compounds released, when using these systems, prevent the toxicity enclosed to peaks and valleys of concentration in target tissues. Therefore, the delivery systems developed resorting to the application of this copolymer can minimize the side effects associated with frequent injections for the administration of therapeutics of interest, without reducing the efficiency of the bioactive compound, serving as a promising platform for avoiding pathophysiological complications [173]. However, the demonstration of such advantages is based on in vitro evidence, and remains to be further elucidated upon in vivo evaluations.

Studies on nanoencapsulation and controlled release from nanostructured carriers have demonstrated the safety of these systems and their contribution to achieve operative concentrations in target cells and tissues, when applying polymer- and lipid-based nanostructured systems [191]. Due to this and according to previous reports, on the capacity of such carriers to protect instable therapeutic proteins, e.g., from enzymatic degradation and other environmental stress factors [191], similar approaches could be applied to the administration of bioactive phytochemicals and take advantage of their biological potential, preventing their metabolization in vivo. In this frame, the research on the beneficial effects of anthocyanins loaded nanoparticles against neurotoxicity in vitro, has been focused on the evaluation of the cytotoxic profile of nanoparticles in the human neuroblastoma model (SH-SY5Y cell line). Upon this study, different concentrations of both free-anthocyanins and anthocyanins loaded in poly(lactide-*co*-clycolide) nanoparticles allowed discarding significant cytotoxic effects, while the joint application of anthocyanins loaded nanoparticles increased the viability of treated cells, by protecting them from neurotoxic events [168]. Thus, poly(lactide-*co*-clycolide) nanoparticles might constitute promising phytochemical carriers with negligible cytotoxicity events. Additionally, the administration of the mentioned nanoparticles provides the advantage of avoiding the need of removing an eventual implant placed as a local source of the bioactive compounds, and thus, no surgery would be required [197,198].

## 5. Toxicity Facts Associated to the Administration of Biodegradable Nanoparticles

In respect to the toxicity associated to the administration of biodegradable nanocarrier systems for bioactive compounds, as far as we are aware, there is no evidence on deleterious effects, although the risks to human health that could be associated to the long term use, remain under explored [148]. the mechanisms responsible for the deleterious effects of nanoparticles in the frame of complex biological systems are associated to increased productions of Reactive Oxygen Species (ROS) and free radicals, that could give rise to oxidative stress, inflammation, and consequently to several disabling pathophysiological situations, associated with solvent residues and polymers toxicity [199]. To shed some light on the actual toxicity of nanoparticles, it is required to evaluate the separate trials independently, paying attention to interactions between nanocarriers and biological systems. This way of processing the available information on the toxicity, associated to nanoparticles, is crucial to establish the reliability of the delivery systems. Actually, to translate successfully nanoformulations from the experimental level to their practical clinical application, it is essential to establish their safety profiles, including the evaluation of immunotoxicity. In this concern, special attention should be paid to the linkage between the physico-chemical and functional properties of nanoparticles (surface charge, size, modulate uptake and interactions with cells, control over surface modification and biodegradation of nanovectors), and the extent in which, these features, contribute to achieve the functional potential envisaged and to minimize potential health risks [200].

Hence, toxicological facts associated to nanocarrier systems involve several physiological, physicochemical and molecular considerations. However, despite the evident interest of understanding the toxicity of these delivery systems, to date, it has not been deeply explored, possibly because the industrial use of this technology is in its infancy. From the information available in the literature regarding this, it is noticed that toxicological facts associated to nanocarriers is closely linked to their size and shape, biomaterials, as well as to their capacity to cross biological barriers [201]. Regarding the molecular mechanisms, behind the toxicological effects, associated to treatments with bioactive compounds carried on biodegradable nanoparticles, it has been noticed the formation of pro-oxidants after their administration, disrupting the balance between their production and the detoxification capacity of cells. This fact also entails augmented inflammation reactions due to the level of redox-sensitive transcription factors [202,203].

In relation with the use of nanoparticles to carry bioactive compounds addressed to be ingested in foods, water or drug delivery devices, the interest of biodegradable nanoparticles and the constraints associated to their toxicity, following oral ingestion, has been associated to the occurrence of deleterious effects on liver, kidney and spleen. In addition, even if the mechanisms responsible for triggering the immune response induced by nanoparticles (especially concerning non-protein-based carriers) are not clear, there is a growing attention on possible allergic reaction [202]. To gain a further insight in this issue, further studies of the immunogenicity of nanocarriers are required to understand under which conditions they are identified by the immune system, deserving the generation of a specific immune responses [202]. Polyethylene glycol (PEG)-grafted liposome infusion has been described to trigger non-IgE-mediated signs of hypersensitivity [204]. On the other hand, advantages can be obtained from the immunogenic characteristics of nanoparticles, as these features suggest them as valuable adjuvants, for instance, for the development of vaccines.

## 6. Future Perspectives for Targeting and Controlled Delivery

In the coming years, the research trends in the application of edible nanoparticles to the administration of bioactive phytochemicals will be closely linked to new manufacturing strategies that combine multiple structural designs, more specific for the diverse types of bioactive compounds. Achieving this objective is foreseen to strengthen the property of nanoparticles and to combine the benefits identified on two or more types of biomaterials. This strategy would contribute to maximize the benefits of loading bioactive compounds on nanoparticles, in terms of effectivity, extended release, and in situ delivery control. Among the desired features, target delivery has not been completely developed yet, remaining an issue with future prospects for implementation. In this sense, materials and fabrication strategies would allow the creation of edible nanoparticles with improved properties is constantly ongoing. Once improved, the specificity and versatility of available nanoparticles, the administration of bioactive compounds orally, by using edible nanoparticles, will constitute a valuable approach that would allow to take advantage from the actual potential of phytochemicals to prevent and treat specific pathophysiological status, due to accurate tissue-targeted delivery.

One of the major constraints enclosed to the administration of bioactive compounds, even when using nanotechnology products, lies in the degradation of these compounds in the gastrointestinal lumen before being absorbed. Thus, although in vitro the works, available in the literature, reveal promising potential of the use of nanoparticles for the administration of phytochemicals loaded nanoparticles, nowadays it is required to complete the determination of the impact of gastrointestinal digestion, by informative in vitro simulation models, as cost-efficient tools for forecasting the oral bioavailability [167,205]. With this aim, in addition to implementing a model mimicking the salivary, gastric, and intestinal fluids concerning salt and enzyme concentrations, it is required to avoid an excessive simplification of the models, by including appropriate simulators of gastrointestinal dynamics, structure, and mechanical issues.

the consideration of gut metabolism, on the actual biological activity of compounds ingested by oral administration, requires a review of the current definition of "bioaccessibility" and "bioavailability", emphasizing the value of the absorption in the upper gastrointestinal tract. Indeed, information on the features of this physiological process, according to the chemical properties of the bioactive compounds of interest, and the derivatives that can be formed during the gastrointestinal digestion, is crucial for the rational design of the specific nanoparticle system. With this objective, the selection of the type of nanoparticles, to improve the absorption at the gastric and small intestine (duodenum) level, may not serve as a gold standard anymore, because, for some phytochemicals, high bioefficacy is associated with the additional compounds synthesized, for instance in the large intestine, as a result of the metabolism of the local microbiota, turning into undesirable their absorption in the upper gastrointestinal.

Besides, when evaluating the effects of nanoparticles and/or nanovesicles on the bioefficacy of phytochemicals, special attention should be paid to eventual changes in the nanoparticle structure as a result of the digestive conditions. Indeed, this modification of the nanoparticles structure could entail changes in their functionality as carriers of bioactive nutrients and non-nutrients and, consequently, induce misunderstanding results. Thus, considering that, when mixed with digestion fluids, the dilution would have a crucial effect on the stability of micelles, liposomes, and nanoemulsions, is required. This deserves to be considered because the surfactant concentration should be maintained in a slim range. Moreover, pH and ions environments in the gastrointestinal fluids, as well as the enzymatic activity in these compartments could also compromise the stability of nanoparticles and the successful development of their role as carriers of bioactive compounds [122]. According to these constraints, it is necessary to evaluate the behavior of nanoparticles in complex matrices and biological systems that will provide actual information on their practical significance.

An additional situation, that needs to be addressed, is the extent in which nanoparticles modify the pharmacokinetics of bioactive phytochemicals, which would contribute to draw new applications, for instance as boosters of the internalization speed. Clarifying this situation is essential for the proper design of the sampling timing to determine pharmacokinetics bioavailability and bioactivity of bioactive compounds subject of study. Overall, encapsulated phytochemicals within appropriate nanoparticles enhance the bioavailability by protecting them from degradation during storage and gastrointestinal digestion, improving solubility in aqueous media, augmenting contact time with the intestinal wall, increasing of the mucus penetration and intestinal permeation, facilitating cellular uptake, prolonging residence time within the body circulation, controlling release rate and site, and altering, according to the envisaged results, the microbiota metabolism. In addition, it should be considered that loading bioactive compounds in nanoparticles, somehow, could be

*Foods* **2018**, *7*, 72

an interesting strategy to prevent their metabolic conversion, while to date, little is known about how the modification of cellular signaling routes differs between free and nanoparticles-linked phytochemicals. In this regard, seems evident that the interaction between phytochemicals and matrix material, or molecules in complex biological systems (for instance in mammals), deserves to be further explored concerning metabolism and bioefficacy [122]. Hence, the information retrieved from studies, focused on the correlation between the application of bioactive compounds carried by nanoparticles with the metabolic profile, biodistribution and bioactivity, will be very useful as a guide for future evaluation of the bioefficacy and predictable in vivo properties.

Concerning the final intestinal stage, the interactions between phytochemicals and drugs with gut microbiota, as well as the consequences of these interactions on bioefficacy, remain underexplored, even if it seems evident that they constitute additional challenges. Indeed, the application of the gathered knowledge on the use of nanoparticles, as carriers of bioactive phytochemicals, would provide the opportunity to decode and take advantage of the reciprocal interactions between phytochemicals and gut microbiota.

Despite the rational utility of the nanotechnology to enhance the bioavailability and bioactivity of bioactive compounds, nanoparticles are synthesized by physical and chemical methods, by using expensive and hazardous chemicals, especially concerning metallic biomaterials, which could limit their actual application in vivo. These constraints are also enclosed to the environmental impact derived of the residues, generated from the nanoparticles synthesis, being required green technologies using no toxic reagents to prepare metal nanoparticles. In this regard, the synthesis of nanoparticles using eco-friendly and biocompatible reagents could contribute to minimize the side effect of these processes.

## 7. Conclusions

Nutraceuticals' potential benefits have been demonstrated, but increasing applications in the near future, for the prevention of diseases onset and severity, are highly demanded. To take maximum advantages of bioactive compounds (nutrients and non-nutrients), responsible for the functionality of nutraceuticals, promising applications have been suggested based on combinations of individual, well characterized, bioactive phytochemicals. This approach would allow synergic situations that would help to reduce dosages and, thus, eventual toxic side effects, while becoming instrumental in the prevention of an array of pathophysiological situations.

To improve the bioavailability of phytochemicals identified as valuable bioactive compounds, in the frame of a specific clinical entity of health disorders, nanoparticles have been investigated as promising delivery systems that could contribute to control release and to fine-tune pharmacokinetics, bioavailability, and bioefficacy. Indeed, nanotechnology would contribute to improve stability of encapsulated bioactive nutrients and non-nutrients against environmental changes (in the diverse in vivo environments) and to release control. Therefore, it should be assumed that nanoparticles have great potential as phytochemical carriers, as, at the dosages monitored, they are not cytotoxic. However, critical aspects and potential for application of nanotechnology in Food Science and Technology field, should be noticed, especially regarding the implementation of green technology for their development and their use in the studies of additional issues, such as the crucial role of the intestinal microbiota for obtaining additional bioactive metabolites, contributing to the health benefits attributed to plant foods and nutraceuticals.

**Author Contributions:** M.M.-B. and R.D.-P. conceived the focus of this review and drafted the manuscript. A.G.-I. and C. G.-V. supervised the project and contributed to the final version of the manuscript upon critical revision of the texts.

**Acknowledgments:** This work was partially funded by the Spanish Ministry of Science and Innovation (MICINN) through the Research Projects AGL2016-75332-C2-1-R, AGL2016-80247-C2-1-R, and AGL2017-83386-R. In addition, this work is included in the framework of the collaboration between the Spanish Research Council (CEBAS-CSIC) and CNRS by "Projets Internationaux de Cooperation Scientifique" (PICS-2015-261141). R.D.-P. was sponsored

*Foods* **2018**, *7*, 72

by a Postdoctoral Contract (Juan de la Cierva de Incorporación ICJI-2015-25373) from the Ministry of Economy, Industry, and Competitiveness of Spain.

**Conflicts of Interest:** the authors declare no conflict of interest.

## Abbreviation

EDC        1-(3-(dimethylamino) propyl) 3-ethylcarbodiimidehydrochloride

## References

1. Pitchaiah, G.; Akula, A.; Parvathaneni, M. Anticancer potential of nutraceutical formulation through antioxidant, anti-inflammatory, and antiproliferative mechanisms in N-methyl-N-nitrosourea-induced mammary cancer. *Int. J. Green Pharm.* **2017**, *11*, 230–235.
2. Henrotin, Y.; Lambert, C.; Couchourel, D.; Ripoll, C.; Chiotelli, E. Nutraceuticals: Do they represent a new era in the management of osteoarthritis?—A narrative review from the lessons taken with five products. *Osteoarthritis Cartil.* **2011**, *19*, 1–21. [CrossRef] [PubMed]
3. Ahmad, A.; Wei, Y.; Syed, F.; Tahir, K.; Taj, R.; Khan, A.U.; Hameed, M.U.; Yuan, Q. Amphotericin B-conjugated biogenic silver nanoparticles as an innovative strategy for fungal infections. *Microb. Pathog.* **2016**, *99*, 271–281. [CrossRef] [PubMed]
4. Mittal, A.K.; Chisti, Y.; Banerjee, U.C. Synthesis of metallic nanoparticles using plant extracts. *Biotechnol. Adv.* **2013**, *31*, 346–356. [CrossRef] [PubMed]
5. Havlickova, B.; Czaika, V.A.; Friedrich, M. Epidemiological trends in skin mycoses worldwide. *Mycoses* **2008**, *51*, 2–15. [CrossRef] [PubMed]
6. Sobel, J.D. Vulvovaginal candidosis. *Lancet* **2007**, *369*, 1961–1971. [CrossRef]
7. Vance, M.E.; Kuiken, T.; Vejerano, E.P.; McGinnis, S.P.; Hochella, M.F., Jr.; Rejeski, D.; Hull, M.S. Nanotechnology in the real world: Redeveloping the nanomaterial consumer products inventory. *Beilstein J. Nanotechnol.* **2015**, *6*, 1769–1780. [CrossRef] [PubMed]
8. Roco, M.C.; Harthorn, B.; Guston, D.; Shapira, P. Innovative and responsible governance of nanotechnology for societal development. In *Nanotechnology Research Directions for Societal Needs in 2020*; Roco, M.C., Mirkin, C.A., Hersam, M.C., Eds.; Springer: Dordrecht, the Netherlands, 2011; pp. 561–617.
9. López-Lorente, A.I.; Mizaikoff, B. Recent advances on the characterization of nanoparticles using infrared spectroscopy. *TrAC Trends Anal. Chem.* **2016**, *84*, 97–106. [CrossRef]
10. Dingman, J. Nanotechnology its impact on food safety. *J. Environ. Health* **2008**, *70*, 47–50. [PubMed]
11. Abbas, K.A.; Saleh, A.M.; Mohamed, A.; MohdAzhan, N. the recent advances in the nanotechnology and its applications in food processing: A review. *J. Food Agric. Environ.* **2009**, *7*, 14–17.
12. Oberdörster, G.; Maynard, A.; Donaldson, K.; Castranova, V.; Fitzpatrick, J.; Ausman, K.; Carter, J.; Karn, B.; Kreyling, W.; Lai, D.; et al. Review Principles for characterizing the potential human health effectsfrom exposure to nanomaterials: Elements of a screening strategy. *Part. Fibre Toxicol.* **2005**, *2*, 8. [CrossRef] [PubMed]
13. Nel, A.; Xia, T.; Mädler, L.; Li, N. Review Toxic Potential of Materials at the Nanolevel. *Science* **2006**, *311*, 622–627. [CrossRef] [PubMed]
14. Khan, I.; Saeed, K.; Khan, I. Nanoparticles: Properties, applications and toxicities. *Arab. J. Chem.* **2017**. [CrossRef]
15. Linsinger, T.P.J.; Chaudhry, Q.; Dehalu, V.; Delahaut, P.; Dudkiewicz, A.; Grombe, R.; von der Kammer, F.; Larsen, E.H.; Legros, S.; Loeschner, K.; et al. Validation of methods for the detection and quantification of engineered nanoparticles in food. *Food Chem.* **2013**, *138*, 1959–1966. [CrossRef] [PubMed]
16. Gallocchio, F.; Bellucoa, S.; Ricci, A. Nanotechnology and food: Brief overview of the current scenario. *Procedia Food Sci.* **2015**, *5*, 85–88. [CrossRef]
17. Rahimnejad, M.; Mokhtarian, N.; Ghasemi, M. Production of protein nanoparticles for food and drug delivery system. *Afr. J. Biotechnol.* **2009**, *8*, 4738–4743.
18. Singh, T.; Shukla, S.; Kumar, P.; Wahla, V.; Bajpai, V.K.; Rather, I.A. Application of Nanotechnology in Food Science: Perception and Overview. *Front. Microbiol.* **2017**, *8*, 1501–1506. [CrossRef] [PubMed]
19. Sant, S.; Tao, S.L.; Fisher, O.Z.; Xu, Q.; Peppas, N.A.; Khademhoseini, A. Microfabrication technologies for oral drug delivery. *Adv. Drug Deliv. Rev.* **2012**, *64*, 496–507. [CrossRef] [PubMed]

20. Momin, J.K.; Jayakumar, C.; Prajapati, J.B. Potential of nanotechnology in functional foods. *Emir. J. Food Agric.* **2013**, *25*, 10–19. [CrossRef]

21. Bengoechea, C.; Jones, O.G.; Guerrero, A.; McClements, D.J. Formation and characterization of lactoferrin/pectin electrostatic complexes: Impact of composition, pH and thermal treatment. *Food Hydrocoll.* **2011**, *25*, 1227–1232. [CrossRef]

22. Berton-Carabin, C.C.; Coupland, J.N.; Elias, R.J. Effect of the lipophilicity of model ingredients on their location and reactivity in emulsions and solid lipid nanoparticles. *Colloids Surf. A Physicochem. Eng. Asp.* **2013**, *431*, 9–17. [CrossRef]

23. Cerqueira, M.A.; Pinheiro, A.C.; Silva, H.D.; Ramos, P.E.; Azevedo, M.A.; Flores-López, M.L.; Rivera, M.C.; Bourbon, A.I.; Ramos, O.L.; Vicente, A.A. Design of bio-nanosystems for oral delivery of functional compounds. *Food Eng. Rev.* **2014**, *6*, 1–19. [CrossRef]

24. Weiss, J.; Takhistov, P.; McClements, D.J. Functional materials in food nanotechnology. *J. Food Sci.* **2006**, *71*, R107–R116. [CrossRef]

25. Neethirajan, S.; Jayas, D.S. Nanotechnology for the food and bioprocessing industries. *Food Bioprocess Technol.* **2011**, *4*, 39–47. [CrossRef]

26. Harde, H.; Das, M.; Jain, S. Solid lipid nanoparticles: An oral bioavailability enhancer vehicle. *Expert Opin. Drug Deliv.* **2011**, *8*, 1407–1424. [CrossRef] [PubMed]

27. Solans, C.; Izquierdo, P.; Nolla, J.; Azemar, N.; Garcia-Celma, M.J. Nano-emulsions. *Curr. Opin. Colloid Interface Sci.* **2005**, *10*, 102–110. [CrossRef]

28. Blasi, P.; Giovagnoli, S.; Schoubben, A.; Ricci, M.; Rossi, C. Solid lipid nanoparticles for targeted brain drug delivery. *Adv. Drug Deliv. Rev.* **2007**, *59*, 454–477. [CrossRef] [PubMed]

29. Thatipamula, R.; Palem, C.; Gannu, R.; Mudragada, S.; Yamsani, M. Formulation and in vitro characterization of domperidone loaded solid lipid nanoparticles and nanostructured lipid carriers. *Daru* **2011**, *19*, 23–32. [PubMed]

30. Gregoriadis, G. Immunological adjuvants: A role for liposomes. *Immunol. Today* **1990**, *11*, 89–97. [CrossRef]

31. Alving, C.R. Liposomes as carriers of antigens and adjuvants. *J. Immunol. Methods* **1991**, *140*, 1–13. [CrossRef]

32. Sharma, A.; Sharma, U.S. Liposomes in drug delivery: Progress and limitations. *Int. J. Pharm.* **1997**, *154*, 123–140. [CrossRef]

33. Vanrooijen, N.; Vannieuwmegen, R. Liposomes in immunology–multilamellar phosphatidylcholine liposomes as a simple, biodegradable and harmless adjuvant without any immunogenic activity of its own. *Immunol. Commun.* **1980**, *9*, 243–256. [CrossRef]

34. Campbell, P.I. Toxicity of some charged lipids used in liposome preparations. *Cytobios* **1983**, *37*, 21–26. [PubMed]

35. Chaize, B.; Colletier, J.P.; Winterhalter, M.; Fournier, D. Encapsulation of enzymes in liposomes: High encapsulation efficiency and control of substrate permeability. *Artif. Cell Blood* **2004**, *32*, 67–75. [CrossRef]

36. Zhou, F.; Neutra, M.R. Antigen delivery to mucosa-associated lymphoid tissues using liposomes as a carrier. *Biosci. Rep.* **2002**, *22*, 355–369. [CrossRef]

37. Matteucci, M.L.; Thrall, D.E. the role of liposomes in drug delivery and diagnostic imaging: A review. *Vet. Radiol. Ultrasound* **2000**, *41*, 100–107. [CrossRef] [PubMed]

38. Mady, M.M.; Ghannam, M.M.; Khalil, W.A.; Repp, R.; Markus, M.; Rascher, W.; Muller, R.; Fahr, A. Efficient gene delivery with serum into human cancer cells using targeted anionic liposomes. *J. Drug Target.* **2004**, *12*, 11–18. [CrossRef] [PubMed]

39. Koo, O.M.; Rubinstein, I.; Onyuksel, H. Role of nanotechnology in targeted drug delivery and imaging: A concise review. *Nanomed. Nanotechnol. Biol. Med.* **2005**, *1*, 193–212. [CrossRef] [PubMed]

40. Jain, R.K. Delivery of molecular and cellular medicine to solid tumors. *J. Control. Release* **1998**, *53*, 49–67. [CrossRef]

41. Kataoka, K.; Harada, A.; Nagasaki, Y. Block copolymer micelles for drug delivery: Design, characterization and biological significance. *Adv. Drug Deliv. Rev.* **2001**, *47*, 113–131. [CrossRef]

42. Gong, J.; Chen, M.; Zheng, Y.; Wang, S.; Wang, Y. Polymeric micelles drug delivery system in oncology. *J. Control. Release* **2012**, *159*, 312–323. [CrossRef] [PubMed]

43. Kwon, G.S.; Okano, T. Polymeric micelles as new drug carriers. *Adv. Drug Deliv. Rev.* **1996**, *21*, 107–116. [CrossRef]

44. Teng, Y.; Morrison, M.E.; Munk, P.; Webber, S.E.; Prochazka, K. Release kinetics studies of aromatic molecules into water from block polymer micelles. *Macromolecules* **1998**, *31*, 3578–3587. [CrossRef]
45. Rapoport, N. Physical stimuli-responsive polymeric micelles for anti-cancer drug delivery. *Prog. Polym. Sci.* **2007**, *32*, 962–990. [CrossRef]
46. Zia, Q.; Farzuddin, M.; Ansari, M.A.; Alam, M.; Ali, A.; Ahmad, A.; Owais, M. Novel drug delivery systems for antifungal compounds. In *Combating Fungal Infections*; Ahmad, I., Owais, M., Shahid, M., Aqil, F., Eds.; Springer: Berlin/Heidelberg, Germany, 2010; pp. 485–528.
47. Mikhail, A.S.; Allen, C. Block copolymer micelles for delivery of cancer therapy: Transport at the whole body, tissue and cellular levels. *J. Control. Release* **2009**, *138*, 214–223. [CrossRef] [PubMed]
48. Malhotra, M.; Jain, N.K. Niosomes as drug carriers. *Indian Drugs* **1994**, *31*, 81–86.
49. Uchegbu, I.F.; Vyas, S.P. Non-ionic surfactant based vesicles (niosomes) in drug delivery. *Int. J. Pharm.* **1998**, *176*, 139–172. [CrossRef]
50. Jain, S.; Singh, P.; Mishra, V.; Vyas, S.P. Mannosylated niosomes as adjuvantcarrier system for oral genetic immunization against Hepatitis B. *Immunol. Lett.* **2005**, *101*, 41–49. [CrossRef] [PubMed]
51. Balasubramaniam, A.; Kumar, V.A.; Pillai, K.S. Formulation and in vivo evaluation of niosome-encapsulated daunorubicin hydrochloride. *Drug Dev. Ind. Pharm.* **2002**, *28*, 1181–1193. [CrossRef] [PubMed]
52. Gude, R.P.; Jadhav, M.G.; Rao, S.G.; Jagtap, A.G. Effects of niosomal cisplatin and combination of the same with theophylline and with activated macrophages in murine B16F10 melanoma model. *Cancer Biother. Radiopharm.* **2002**, *17*, 183–192. [CrossRef] [PubMed]
53. Shahiwala, A.; Misra, A. Studies in topical application of niosomally entrapped Nimesulide. *J. Pharm. Pharm. Sci.* **2002**, *5*, 220–225. [PubMed]
54. Pardeshi, C.; Rajput, P.; Belgamwar, V.; Tekade, A.; Patil, G.; Chaudhary, K.; Sonje, A. Solid lipid based nanocarriers: An overview. *Acta Pharm.* **2012**, *62*, 433–472. [CrossRef] [PubMed]
55. Mukherjee, S.; Ray, S.; Thakur, R.S. Solid Lipid Nanoparticles: A Modern Formulation Approach in Drug Delivery System. *Indian J. Pharm. Sci.* **2009**, *71*, 349–358. [CrossRef] [PubMed]
56. Naseri, N.; Valizadeh, H.; Zakeri-Milani, P. Solid lipid nanoparticles and nanostructured lipid carriers: Structure preparation and application. *Adv. Pharm. Bull.* **2015**, *5*, 305–313. [CrossRef] [PubMed]
57. Singh, R.; Lillard, J.W. Nanoparticle-based targeted drug delivery. *Exp. Mol. Pathol.* **2009**, *86*, 215–223. [CrossRef] [PubMed]
58. Nitta, S.; Numata, K. Biopolymer-based nanoparticles for drug/gene delivery and tissue engineering. *Int. J. Mol. Sci.* **2013**, *14*, 1629–1654. [CrossRef] [PubMed]
59. Fojan, P.; Schwach-Abdellaoui, K.; Tommeraas, K.; Gurevich, L.; Petersen, S.B. Polysaccharide based Nanoparticles and Nanoporous matrices. *Nano Sci. Technol. Inst.* **2006**, *2*, 79–82.
60. Lemarchand, C.; Gref, R.; Couvreur, P. Polysaccharide-decorated nanoparticles. *Eur. J. Pharm. Biopharm.* **2004**, *58*, 327–341. [CrossRef] [PubMed]
61. Liu, Z.; Jiao, Y.; Wang, Y.; Zhou, C.; Zhang, Z. Polysaccharides-based nanoparticles as drug delivery systems. *Adv. Drug Deliv. Rev.* **2008**, *60*, 1650–1662. [CrossRef] [PubMed]
62. Salatin, S.; Jelvehgarim, M. Natural Polysaccharide based Nanoparticles for Drug/Gene Delivery. *Pharm. Sci.* **2017**, *23*, 84–94. [CrossRef]
63. Amidi, M.; Mastrobattista, E.; Jiskoot, W.; Hennink, W.E. Chitosan-based delivery systems for protein therapeutics and antigens. *Adv. Drug Deliv. Rev.* **2010**, *62*, 59–82. [CrossRef] [PubMed]
64. Duceppe, N.; Tabrizian, M. Advances in using chitosan-based nanoparticles for in vitro and in vivo drug and gene delivery. *Expert Opin. Drug Deliv.* **2010**, *7*, 1191–1207. [CrossRef] [PubMed]
65. You, J.O.; Peng, C.A. Calcium-Alginate Nanoparticles Formed by Reverse Microemulsion as Gene Carriers. *Macromol. Symp.* **2005**, *219*, 147–153. [CrossRef]
66. Ojea-Jiménez, I.; Tort, O.; Lorenzo, J.; Puntes, V.F. Engineered nonviral nanocarriers for intracellular gene delivery applications. *Biomed. Mater.* **2012**, *7*, 1–6. [CrossRef] [PubMed]
67. Sun, J.; Tan, H. Alginate-based biomaterials for regenerative medicine applications. *Materials* **2013**, *6*, 1285–1309. [CrossRef] [PubMed]
68. Arpicco, S.; Milla, P.; Stella, B.; Dosio, F. Hyaluronic Acid Conjugates as Vectors for the Active Targeting of Drugs, Genes and Nanocomposites in Cancer Treatment. *Molecules* **2014**, *19*, 3193–3230. [CrossRef] [PubMed]

69. Zhu, S.F.; Zheng, J.; Liu, F.; Qiu, C.Y.; Lin, W.F.; Tang, C.H. the influence of ionic strength on the characteristics of heat-induced soy protein aggregate nanoparticles and the freeze–thaw stability of the resultant Pickering emulsions. *Food Funct.* **2017**, *8*, 2974–2981. [CrossRef] [PubMed]

70. Tarhini, M.; Greige-Gerges, H.; Elaissari, A. Protein-based nanoparticles: From preparation to encapsulation of active molecules. *Int. J. Pharm.* **2017**, *522*, 172–197. [CrossRef] [PubMed]

71. Kaul, G.; Amiji, M. Biodistribution and targeting potential of poly(ethylene glycol)-modified gelatin nanoparticles in subcutaneous murine tumor model. *J. Drug Target.* **2004**, *12*, 585–591. [CrossRef] [PubMed]

72. Donsìa, F.; Sessaa, M.; Mediounic, H.; Mgaidic, A.; Ferrari, G. Encapsulation of bioactive compounds in nanoemulsionbased delivery systems. *Procedia Food Sci.* **2011**, *1*, 1666–1671. [CrossRef]

73. Ahmed, K.; Li, Y.; McClements, D.J.; Xiao, H. Nanoemulsion- and emulsion-based delivery systems for curcumin: Encapsulation and release properties. *Food Chem.* **2012**, *132*, 799–807. [CrossRef]

74. Schultz, S.; Wagner, G.; Urban, K.; Ulrich, J. High-pressure homogenization as a process for emulsion formation. *Chem. Eng. Technol.* **2004**, *27*, 361–368. [CrossRef]

75. McClements, D.J. *Food Emulsions: Principles, Practice and Techniques*; CRC Press: Boca Raton, FL, USA, 2005.

76. Montes de Oca-Ávalos, J.M.; Herrera, M.L. Nanoemulsions: Stability and physical properties. *Curr. Opin. Food Sci.* **2017**, *16*, 1–6. [CrossRef]

77. Aboalnaja, K.O.; Yaghmoor, S.; Kumosani, T.A.; McClements, D.J. Utilization of nanoemulsions to enhance bioactivity of pharmaceuticals, supplements, and nutraceuticals: Nanoemulsion delivery systems and nanoemulsion excipient systems. *Expert Opin. Drug Deliv.* **2016**, *13*, 1–10. [CrossRef] [PubMed]

78. Kumar, K.; Sarkar, P. Encapsulation of bioactive compounds using nanoemulsions. *Environ. Chem. Lett.* **2018**, *16*, 59–70. [CrossRef]

79. Li, M.; Cui, J.; Ngadi, M.O.; Ma, Y. Absorption mechanism of whey-protein-delivered curcumin using caco-2 cell monolayers. *Food Chem.* **2015**, *180*, 48–54. [CrossRef] [PubMed]

80. Yu, H.; Huang, Q. Investigation of the cytotoxicity of food-grade nanoemulsions in caco-2 cell monolayers and hepg2 cells. *Food Chem.* **2013**, *141*, 29–33. [CrossRef] [PubMed]

81. Sari, T.P.; Mann, B.; Kumar, R.; Singh, R.R.B.; Sharma, R.; Bhardwaj, M.; Athira, S. Preparation and characterization of nanoemulsion encapsulating curcumin. *Food Hydrocoll.* **2015**, *43*, 540–546. [CrossRef]

82. Sessa, M.; Tsao, R.; Liu, R.; Ferrari, G.; Donsì, F. Evaluation of the stability and antioxidant activity of nanoencapsulated resveratrol during in vitro digestion. *J. Agric. Food Chem.* **2011**, *59*, 12352–12360. [CrossRef] [PubMed]

83. Sessa, M.; Balestrieri, M.L.; Ferrari, G.; Servillo, L.; Castaldo, D.; D'Onofrio, N.; Donsì, F.; Tsao, R. Bioavailability of encapsulated resveratrol into nanoemulsion-based delivery systems. *Food Chem.* **2014**, *147*, 42–50. [CrossRef] [PubMed]

84. Wenzel, E.; Somoza, V. Metabolism and bioavailability of trans-resveratrol. *Mol. Nutr. Food Res.* **2005**, *49*, 472–481. [CrossRef] [PubMed]

85. Gupta, A.; Eral, H.B.B.; Hatton, T.A.; Doyle, P.S. Controlling and predicting droplet size of nanoemulsions: Scaling relations with experimental validation. *Soft Matter* **2016**, *12*, 1452–1458. [CrossRef] [PubMed]

86. Zeeb, B.; Herz, E.; McClements, D.J.; Weiss, J. Reprint of: Impact of alcohols on the formation and stability of protein-stabilized nanoemulsions. *J. Colloid Interface Sci.* **2015**, *449*, 13–20. [CrossRef] [PubMed]

87. Wang, T.; Soyama, S.; Luo, Y. Development of a novel functional drink from all natural ingredients using nanotechnology. *LWT Food Sci. Technol.* **2016**, *73*, 458–466. [CrossRef]

88. Pan, L.; Liu, J.; He, Q.; Shi, J. MSN-mediated sequential valscular to-cell nuclear targeted drug delivery for efficient tumor regression. *Adv. Mater.* **2014**, *26*, 6742–6748. [CrossRef] [PubMed]

89. Sarkar, S.; Siddiqui, A.A.; Mazumder, S.; De, R.; Saha, S.J.; Banerjee, C.; Iqbal, M.S.; Adhikari, S.; Alam, A.; Roy, S.; et al. Elagic Acid, a Dietary Polyphenol, Unhibits Tautomerase Activity of Human Magrophage Migration Inhibitory Factor and Its Pro-Inflammatory Responses in Human Perfipheral Blood Mononuclear Cells. *J. Agric. Food Chem.* **2015**, *63*, 4988–4998. [CrossRef] [PubMed]

90. Ezhilarasi, P.N.; Karthik, P.; Chhanwal, N.; Anandharamakrishnan, C. Nanoencapsulation techniques for food bioactive components: A review. *Food Bioprocess Technol.* **2013**, *6*, 628–647. [CrossRef]

91. Fathi, M.; Martín, T.; McClements, D.J. Nanoencapsulation of food ingredients using acrobhidrate based delivery systems. *Trends Food Sci. Technol.* **2014**, *39*, 19–38. [CrossRef]

92. Shani-Levi, C.; Levi-Tal, S.; Lesmes, U. Comparative performance of mild proteins and their emulsions under dynamic in vitro adult and infant gastric digestion. *Food Hydrocoll.* **2013**, *32*, 349–357. [CrossRef]

93. Wang, X.; Jiang, Y.; Wang, Y.W.; Huang, M.T.; Ho, C.T.; Huang, Q. Enhancing anti-inflammation activity of curcumin through O/W nanoemulsions. *Food Chem.* **2008**, *108*, 419–424. [CrossRef] [PubMed]
94. Cheng, A.L.; Hsu, C.H.; Lin, J.K.; Hsu, M.M.; Ho, Y.F.; Shen, T.S.; Ko, J.Y.; Lin, J.T.; Lin, B.R.; Ming-Shiang, W.; et al. Phase I clinical trial of curcumin, a chemopreventive agent, in patients with high-risk or pre-malignant lesions. *Anticancer Res.* **2001**, *21*, 2895–2900. [PubMed]
95. Simion, V.; Stan, D.; Constantinescu, C.A.; Deleanu, M.; Dragan, E.; Tucureanu, M.M.; Gan, A.M.; Butoi, E.; Constantin, A.M.; Maduteanu, I.; et al. Conjugation of curcumin-loaded lipid nanoemulsions with cell-penetrating peptides increases their cellular uptake and enhances the anti-inflammatory effects in endothelial cells. *J. Pharm. Pharmacol.* **2016**, *68*, 195–207. [CrossRef] [PubMed]
96. Yu, H.; Huang, Q. Enhanced in vitro anti-cancer activity of curcumin encapsulated in hydrophobically modified starch. *Food Chem.* **2010**, *119*, 669–674. [CrossRef]
97. Bourassa, P.; Kanakis, C.D.; Tarantilis, P.; Pollissiou, M.G.; Tajmir-Riahi, H.A. Resveratrol, genistein, and curcumin bind bovine serum albumin. *J. Phys. Chem. B* **2010**, *114*, 3348–3354. [CrossRef] [PubMed]
98. Mohammadi, F.; Bordbar, A.K.; Divsalar, A.; Mohammadi, K.; Saboury, A. Interaction of curcumin and acetylcurcumin with the lipocalin member beta-lactoglobulin. *Protein J.* **2009**, *28*, 117–123. [CrossRef] [PubMed]
99. Shelma, R.; Sharma, C.P. Acyl modified chitosan derivatives for oral delivery of insulin and curcumin. *J. Mater. Sci. Mater. Med.* **2010**, *21*, 2133–2140. [CrossRef] [PubMed]
100. Scalbert, A.; Williamson, G. Dietary intake and bioavailability of polyphenols. *J. Nutr.* **2000**, *130*, 2073S–2085S. [CrossRef] [PubMed]
101. Koutelidakis, A.E.; Argyri, K.; Sevastou, Z.; Lamprinaki, D.; Panagopoulou, E.; Paximada, E.; Sali, A.; Papalazarou, V.; Mallouchos, A.; Evageliou, V.; et al. Bioactivity of Epigallocatechin Gallate Nanoemulsions Evaluated in Mice Model. *J. Med. Food* **2017**, *20*, 923–931. [CrossRef] [PubMed]
102. Orafidiya, L.O.; Agbani, E.O.; Oyedele, A.O.; Babalola, O.O.; Onayemi, O. Preliminary clinical tests on topical preparations of *Ocimum gratissimum* linn leaf essential oil for the treatment of Acne vulgaris. *Clin. Drug Investig.* **2002**, *22*, 313–319. [CrossRef]
103. Baranauskiene, R.M.; Venskutonis, P.R.; Dewettinck, K.; Verhe, R. Properties of oregano (*Origanum vulgare* L.), citronella (*Cymbopogon nardus* G.) and marjoram (*Majorana hortensis* L.) flavors encapsulated into milk protein-based matrices. *Food Res. Int.* **2006**, *39*, 413–425. [CrossRef]
104. Wang, X.; Wang, Y.W.; Huang, Q. Enhancing stability and oral bioavailability of polyphenols using nanoemulsions. *ACS Symp. Ser.* **2009**, *1007*, 198–212.
105. Bilia, A.R.; Guccione, C.; Isacchi, B.; Righeschi, C.; Firenzuoli, F.; Bergonzi, M.C. Essential Oils Loaded in Nanosystems: A Developing Strategy for a Successful Therapeutic Approach. *Evid.-Based Complement. Altern. Med.* **2014**, *2014*, 1–14. [CrossRef] [PubMed]
106. Maryam, I.; Huzaifa, U.; Hindatu, H.; Zubaida, S. Nanoencapsulation of essential oils with enhanced antimicrobial activity: A new way of combating antimicrobial Resistance. *J. Pharmacogn. Phytochem.* **2015**, *4*, 165–170.
107. Oejhlke, K.; Behsnilian, D.; Mayer-Miebach, E.; Weidler, P.G.; Greiner, R. Edible solid lipid nanopartilces (SLN) as carrier systems for antioxidants of different lipophylicity. *PLoS ONE* **2017**, *12*, e0171662.
108. Jung, S.; Choi, C.H.; Lee, C.S.; Yi, H. Integrated fabrication–conjugation methods for polymeric and hybrid microparticles for programmable drug delivery and biosensing application. *Biotechnol. J.* **2016**, *11*, 1561–1571. [CrossRef] [PubMed]
109. González, M.J.; Medina, I.; Maldonado, O.S.; Lucas, R.; Morales, J.C. Antioxidant activity of alkyl gallates and glycosyl alkyl gallates in fish oil in water emulsions: Relevance of their surface active properties and of thetype of emulsifier. *Food Chem.* **2015**, *183*, 190–196. [CrossRef] [PubMed]
110. Nagao, A.; Kotake-Nara, E.; Hase, M. Effects of fats and oils on the bioaccessibility of carotenoids and vitamin E in vegetables. *Biosci. Biotechnol. Biochem.* **2013**, *77*, 776–785. [CrossRef] [PubMed]
111. Vidhyalakshmi, R.; Bhakyaraj, R.; Subhasree, R. Encapsulation "the future of probiotics"—A review. *Adv. Biol. Res.* **2009**, *3*, 96–103.
112. Hou, R.C.W.; Lin, M.Y.; Wang, M.M.C.; Tzen, J.T.C. Increase of viability of entrapped cells of *Lactobacillus delbrueckii* ssp. *bulgaricus in artificial sesame oils emulsions. J. Dairy Sci.* **2003**, *86*, 424–428. [PubMed]

113. Kloover, J.S.; den Bakker, M.A.; Gelderblom, H.; van Meerbeeck, J.P. Fatal outcome of a hypersensitivity reaction to paclitaxel: A critical review of premedication regimens. *Br. J. Cancer* **2004**, *90*, 304–305. [CrossRef] [PubMed]

114. Rijcken, C.J.F.; Soga, O.; Hennink, W.E.; van Nostrum, C.F. Triggered destabilisation of polymeric micelles and vesicles by changing polymers polarity: An attractive tool for drug delivery. *J. Control. Release* **2007**, *120*, 131–148. [CrossRef] [PubMed]

115. Chen, Y.C.; Lo, C.L.; Lin, Y.F.; Hsiue, G.H. Rapamycin encapsulated in dualresponsive micelles for cancer therapy. *Biomaterials* **2013**, *34*, 1115–1127. [CrossRef] [PubMed]

116. Matsumura, Y.; Kataoka, K. Preclinical and clinical studies of anticancer agent-incorporating polymer micelles. *Cancer Sci.* **2009**, *100*, 572–579. [CrossRef] [PubMed]

117. Swamy, M.K.; Sinniah, U.R. A comprehensive review on the phytochemical constituents and pharmacological activities of pogostemon cablin benth: An aromatic medicinal plant of industrial importance. *Molecules* **2015**, *20*, 8521–8547. [CrossRef] [PubMed]

118. Pinciroli, M.; Domínguez-Perles, R.; Abellán, A.; Guy, A.; Durand, T.; Oger, C.; Galano, J.M.; Ferreres, F.; Gil-Izquierdo, A. Comparative study of the Phytoprostane and Phytofuran Content of indica and japonica Rice (*Oryza sativa* L.) Flours. *J. Agric. Food Chem.* **2017**, *65*, 8938–8947. [CrossRef] [PubMed]

119. Lu, B.Y.; Li, M.Q.; Yin, R. Phytochemical Content, Health benefits, and toxicology of common edible flowers: A review (2000–2015). *Crit. Rev. Food Sci. Nutr.* **2016**, *56*, S130–S148. [CrossRef] [PubMed]

120. Chang, S.K.; Alasalvar, C.; Shahidi, F. Review of dried fruits: Phytochemicals, antioxidant efficacies, and health benefits. *J. Funct. Foods* **2016**, *21*, 113–132. [CrossRef]

121. Son, Y.R.; Choi, E.H.; Kim, G.T.; Park, T.S.; Shim, S.M. Bioefficacy of graviola leaf extracts in scavenging free radicals and upregulating antioxidant genes. *Food Funct.* **2016**, *7*, 861–871. [CrossRef] [PubMed]

122. Xiao, J.; Cao, Y.; Huang, Q. Edible Nanoencapsulation Vehicles for Oral Delivery of Phytochemicals: A Perspective Paper. *J. Agric. Food Chem.* **2017**, *65*, 6727–6735. [CrossRef] [PubMed]

123. Medina, S.; Domínguez-Perles, R.; Gil, J.I.; Ferreres, F.; Gil-Izquierdo, A. Metabolomics and the Diagnosis of Human Diseases—A Guide to the Markers and Pathophysiological Pathways Affected. *Curr. Med. Chem.* **2014**, *21*, 823–848. [CrossRef] [PubMed]

124. Ting, Y.; Jiang, Y.; Ho, C.T.; Huang, Q. Common delivery systems for enhancing in vivo bioavailability and biological efficacy of nutraceuticals. *J. Funct. Foods* **2014**, *7*, 112–128. [CrossRef]

125. Melo-Filho, C.C.; Braga, R.C.; Andrade, C.H. Advances in methods for predicting phase I metabolism of polyphenols. *Curr. Drug Metab.* **2014**, *15*, 120–126. [CrossRef] [PubMed]

126. Huang, Q.R.; Yu, H.L.; Ru, Q.M. Bioavailability and delivery of nutraceuticals using nanotechnology. *J. Food Sci.* **2010**, *75*, R50–R57. [CrossRef] [PubMed]

127. Radhakrishnan, R.; Kulhari, H.; Pooja, D.; Gudem, S.; Bhargava, S.; Shukla, R.; Sistla, R. Encapsulation of biophenolic phytochemical EGCG within lipid nanoparticles enhances its stability and cytotoxicity against cancer. *Chem. Phys. Lipids* **2016**, *198*, 51–60. [CrossRef] [PubMed]

128. Lockman, P.R.; Koziara, J.M.; Mumper, R.J.; Allen, D.D. Nanoparticle surface charges alter blood-brain barrier integrity and permeability. *J. Drug Target.* **2004**, *12*, 635–641. [CrossRef] [PubMed]

129. Liu, Y.; Li, K.; Liu, B.; Feng, S.S. A strategy for precision engineering of nanoparticles of biodegradable copolymers for quantitative control of targeted drug delivery. *Biomaterials* **2010**, *31*, 9145–9155. [CrossRef] [PubMed]

130. Kumari, A.; Singla, R.; Guliani, A.; Yadav, S.K. Nanoencapsulation for drug delivery. *EXCLI J.* **2014**, *13*, 265–286. [PubMed]

131. Xiao, J.; Nian, S.; Huang, Q.R. Assembly of kafirin/carboxymethyl chitosan nanoparticles to enhance the cellular uptake of curcumin. *Food Hydrocoll.* **2015**, *51*, 166–175. [CrossRef]

132. Kumar, D.V.; Verma, P.R.P.; Singh, S.K. Development and evaluation of biodegradable polymeric nanoparticles for the effective delivery of quercetin using a quality by design approach. *LWT Food Sci. Technol.* **2015**, *61*, 330–338. [CrossRef]

133. Pan, K.; Zhong, Q.X.; Baek, S.J. Enhanced dispersibility and bioactivity of curcumin by encapsulation in casein nanocapsules. *J. Agric. Food Chem.* **2013**, *61*, 6036–6043. [CrossRef] [PubMed]

134. Hu, B.; Ting, Y.; Yang, X.; Tang, W.; Zeng, X.; Huang, Q. Nanochemoprevention by encapsulation of (−)-epigallocatechin-3gallate with bioactive peptides/chitosan nanoparticles for enhancement of its bioavailability. *Chem. Commun.* **2012**, *48*, 2421–2423. [CrossRef] [PubMed]

135. Zou, L.Q.; Zheng, B.J.; Zhang, R.J.; Zhang, Z.P.; Liu, W.; Liu, C.M.; Xiao, H.; McClements, D.J. Food-grade nanoparticles for encapsulation, protection and delivery of curcumin: Comparison of lipid, protein, and phospholipid nanoparticles under simulated gastrointestinal conditions. *RSC Adv.* **2016**, *6*, 3126–3136. [CrossRef]

136. Domínguez-Perles, R.; Abellán, A.; León, D.; Ferreres, F.; Guy, A.; Oger, C.; Galano, J.M.; Durand, T.; Gil-Izquierdo, A. sorting out the phytoprostane and phytofuran profile in vegetable oils. *Food Res. Int.* **2018**, *107*, 619–628. [CrossRef] [PubMed]

137. Xiao, J.; Li, C.; Huang, Q. Kafirin nanoparticles-stabilized Pickering emulsions as oral delivery vehicles: Physicochemical stability and in vitro digestion profile. *J. Agric. Food Chem.* **2015**, *63*, 10263–10270. [CrossRef] [PubMed]

138. Liu, M.; Zhang, J.; Zhu, X.; Shan, W.; Li, L.; Zhong, J.J.; Zhang, Z.R.; Huang, Y. Efficient mucus permeation and tight junction opening by dissociable "mucus-inert" agent coated trimethyl chitosan nanoparticles for oral insulin delivery. *J. Control. Release* **2016**, *222*, 67–77. [CrossRef] [PubMed]

139. Perez, Y.A.; Urista, C.M.; Martinez, J.I.; Nava, M.D.D.; Rodriguez, F.A.R. Functionalized polymers for enhance oral bioavailability of sensitive molecules. *Polymers* **2016**, *8*, 214. [CrossRef]

140. Barua, S.; Mitragotri, S. Challenges associated with penetration of nanoparticles across cell and tissue barriers: A review of current status and future prospects. *Nano Today* **2014**, *9*, 223–243. [CrossRef] [PubMed]

141. Johnson, B.M.; Charman, W.N.; Porter, C.J.H. the impact of P-glycoprotein efflux on enterocyte residence time and enterocytebased metabolism of verapamil. *J. Pharm. Pharmacol.* **2001**, *53*, 1611–1619. [CrossRef] [PubMed]

142. Charman, W.N.A.; Noguchi, T.; Stella, V.J. An experimental system designed to study the in situ intestinal lymphatic transport of lipophilic drugs in anesthetized rats. *Int. J. Pharm.* **1986**, *33*, 155–164. [CrossRef]

143. Kim, H.; Kim, Y.; Lee, J. Liposomal formulations for enhanced lymphatic drug delivery. *Asian J. Pharm. Sci.* **2013**, *8*, 96–103. [CrossRef]

144. Sha, X.; Yan, G.; Wu, Y.; Li, J.; Fang, X. Effect of selfmicroemulsifying drug delivery systems containing Labrasol on tight junctions in Caco-2 cells. *Eur. J. Pharm. Sci.* **2005**, *24*, 477–486. [CrossRef] [PubMed]

145. Xu, X.F.; Xu, P.P.; Ma, C.W.; Tang, J.; Zhang, X.W. Gut microbiota, host health, and polysaccharides. *Biotechnol. Adv.* **2013**, *31*, 318–337. [CrossRef] [PubMed]

146. Caesar, R.; Nygren, H.; Oresic, M.; Backhed, F. Interaction between dietary lipids and gut microbiota regulates hepatic cholesterol metabolism. *J. Lipid Res.* **2016**, *57*, 474–481. [CrossRef] [PubMed]

147. Aqil, F.; Munagala, R.; Jeyabalan, J.; Vadhanam, M.V. Bioavailability of phytochemicals and its enhancement by drug delivery systems. *Cancer Lett.* **2013**, *334*, 133–141. [CrossRef] [PubMed]

148. Kumari, A.; Yadav, S.K.; Yadav, S.C. Biodegradable polymeric nanoparticles based drug delivery systems. *Colloids Surf. B Biointerfaces* **2010**, *75*, 1–18. [CrossRef] [PubMed]

149. Boulle, F.; Kenis, G.; Cazorla, M.; Hamon, M.; Steinbusch, H.W.; Lanfumey, L.; van del Hove, D.L. TrkB inhibition as a therapeutic target for CNS-related disorders. *Prog. Neurobiol.* **2012**, *98*, 197–206. [CrossRef] [PubMed]

150. Huang, E.J.; Reichardt, L.F. Neurotrophins: Roles in neuronal development and function. *Annu. Rev. Neurosci.* **2001**, *24*, 677–736. [CrossRef] [PubMed]

151. Cowansage, K.K.; LeDoux, J.E.; Monfils, M.H. Brain-derived neurotrophic factor: A dynamic gatekeeper of neural plasticity. *Curr. Mol. Pharmacol.* **2010**, *3*, 12–29. [CrossRef] [PubMed]

152. Lindholm, P.; Voutilainen, M.H.; Laurén, J.; Peränen, J.; Leppänen, V.M.; Andressoo, J.O.; Lindahl, M.; Janhunen, S.; Kalkkinen, N.; Timmusk, T.; et al. Novel neurotrophic factor CDNF protects and rescues midbrain dopamine neurons in vivo. *Nature* **2007**, *448*, 73–77. [CrossRef] [PubMed]

153. Pattarawarapan, M.; Burgess, K. Molecular basis of neurotrophin-receptor interactions. *J. Med. Chem.* **2003**, *46*, 5277–5291. [CrossRef] [PubMed]

154. Gullett, N.P.; Ruhul Amin, A.R.; Bayraktar, S.; Pezzuto, J.M.; Shin, D.M.; Khuri, F.R.; Aggarwal, B.B.; Surh, Y.J.; Kucuk, O. Cancer prevention with natural compounds. *Semin. Oncol.* **2010**, *37*, 258–281. [CrossRef] [PubMed]

155. Naithani, R.; Huma, L.C.; Moriarty, R.M.; McCormick, D.L.; Mehta, R.G. Comprehensive review of cancer chemopreventive agents evaluated in experimental carcinogenesis models and clinical trials. *Curr. Med. Chem.* **2008**, *15*, 1044–1071. [CrossRef] [PubMed]

156. Zolk, O.; Fromm, M.F. Transporter-mediated drug uptake and efflux: Important determinants of adverse drug reactions. *Clin. Pharmacol. Ther.* **2011**, *89*, 798–805. [CrossRef] [PubMed]

157. Kusuhara, H.; Sugiyama, Y. Role of transporters in the tissue-selective distribution and elimination of drugs: Transporters in the liver, small intestine, brain and kidney. *J. Control. Release* **2002**, *78*, 43–54. [CrossRef]

158. Li, Y.; Paxton, J.W. Oral bioavailability and disposition of phytochemicals. In *Phytochemicals—Bioactivities and Impact on Health*; Rasooli, I., Ed.; InTech: Rijeka, Croatia, 2011; pp. 117–138.

159. Manach, C.; Scalbert, A.; Morand, C.; Remesy, C.; Jimenez, L. Polyphenols: Foodsources and bioavailability. *Am. J. Clin. Nutr.* **2004**, *79*, 727–747. [CrossRef] [PubMed]

160. Mishra, B.; Patel, B.B.; Tiwari, S. Colloidal nanocarriers: A review on formulation technology, types and applications toward targeted drug delivery. *Nanomed. Nanotechnol.* **2010**, *6*, 9–24. [CrossRef] [PubMed]

161. Oerlemans, C.; Bult, W.; Bos, M.; Storm, G.; Nijsen, J.F.W.; Hennink, W.E. Polymeric micelles in anticancer therapy: Targeting, imaging and triggered release. *Pharm. Res.* **2010**, *27*, 2569–2589. [CrossRef] [PubMed]

162. Davis, S.S. Biomedical applications of nanotechnology—Implications for drug targeting and gene therapy. *Trends Biotechnol.* **1997**, *15*, 217–224. [CrossRef]

163. Zhao, Y. Photocontrollable block copolymer micelles: What can we control? *J. Mater. Chem.* **2009**, *19*, 4887–4895. [CrossRef]

164. Gil, E.S.; Hudson, S.M. Stimuli-reponsive polymers and their bioconjugates. *Prog. Polym. Sci.* **2004**, *29*, 1173–1222. [CrossRef]

165. Shenoy, D.; Little, S.; Langer, R.; Amiji, M. Poly(ethylene oxide)-modified poly(beta-amino ester) nanoparticles as a pH-sensitive system for tumor targeted delivery of hydrophobic drugs: Part 2. In vivo distribution and tumor localization studies. *Pharm. Res.* **2005**, *22*, 2107–2114. [CrossRef] [PubMed]

166. Devalapally, H.; Shenoy, D.; Little, S.; Langer, R.; Amiji, M. Poly(ethylene oxide)-modified poly(beta-amino ester) nanoparticles as a pH-sensitive system for tumor-targeted delivery of hydrophobic drugs: Part 3. Therapeutic efficacy and safety studies in ovarian cancer xenograft model. *Cancer Chemother. Pharm.* **2007**, *59*, 477–484. [CrossRef] [PubMed]

167. Allen, C.; Maysinger, D.; Eisenberg, A. Nano-engineering block copolymer aggregates for drug delivery. *Colloids Surf. B Biointerfaces* **1999**, *16*, 3–27. [CrossRef]

168. Amin, F.U.; Shah, S.A.; Badshah, H.; Khan, M.; Kim, M.O. Anthocyanins encapsulated by PLGA@PEG nanoparticles potentially improved its free radical scavenging capabilities via p38/JNK pathway against Aβ1–42-induced oxidative stress. *J. Nanobiotechnol.* **2017**, *15*, 12–27. [CrossRef] [PubMed]

169. Torchilin, V.P.; Lukyanov, A.N. Peptide and protein drug delivery to and into tumors: Challenges and solutions. *Drug Discov. Today* **2003**, *8*, 259–266. [CrossRef]

170. Jiménez-Aguilar, D.M.; Ortega-Regules, A.E.; Lozada-Ramírez, J.D.; Pérez-Pérez, M.C.I.; Vernon-Cartere, E.J.; Welti-Chanesa, J. Color and chemical stability of spray-dried blueberry extract using mesquite gum as wall material. *J. Food Compos. Anal.* **2011**, *24*, 889–894. [CrossRef]

171. Sampat, K.M.; Garg, S.J. Complications of intravitreal injections. *Curr. Opin. Ophthalmol.* **2010**, *21*, 178–183. [CrossRef] [PubMed]

172. Thakur, S.S.; Barnett, N.L.; Donaldson, M.J.; Parekh, H.S. Intravitreal drug delivery in retinal disease: Are we out of our depth? *Expert Opin. Drug Deliv.* **2014**, *11*, 1575–1590. [CrossRef] [PubMed]

173. Agrahari, V.; Agrahari, V.; Hung, W.T.; Christenson, L.K.; Mitra, A.K. Composite Nanoformulation Therapeutics for Long-Term Ocular Delivery of Macromolecules. *Mol. Pharm.* **2016**, *13*, 2912–2922. [CrossRef] [PubMed]

174. Traka, M.H.; Mithen, R.F. Plant Science and Human Nutrition: Challenges in Assessing Health-Promoting Properties of Phytochemicals. *Plant Cell* **2011**, *23*, 2483–2497. [CrossRef] [PubMed]

175. Skrovankova, S.; Sumczynski, D.; Mlcek, J.; Jurikova, T.; Sochor, J. Bioactive Compounds and Antioxidant Activity in Different Types of Berries. *Int. J. Mol. Sci.* **2015**, *16*, 24673–24706. [CrossRef] [PubMed]

176. Sahoo, S.K.; Dilnawaz, F.; Krishnakumar, S. Nanotechnology in ocular drug delivery. *Drug Discov. Today* **2008**, *13*, 144–151. [CrossRef] [PubMed]

177. Bangham, A.D. Liposomes: the Babraham connection. *Chem. Phys. Lipids* **1993**, *64*, 275–285. [CrossRef]

178. Chiappetta, D.A.; Sosnik, A. Poly (ethylene oxide)-poly (propylene oxide) block copolymer micelles as drug delivery agents: Improved hydrosolubility, stability and bioavailability of drugs. *Eur. J. Pharm. Biopharm.* **2007**, *66*, 303–317. [CrossRef] [PubMed]

179. Labhasetwar, V.; Song, C.; Levy, R.J. Nanoparticle drug delivery system for restenosis. *Adv. Drug Deliv. Rev.* **1997**, *24*, 63–85. [CrossRef]

180. Pescina, S.; Sonvico, F.; Santi, P.; Nicoli, S. Therapeutics and carriers: the dual role of proteins in nanoparticles for ocular delivery. *Curr. Top. Med. Chem.* **2015**, *15*, 369–385. [CrossRef] [PubMed]

181. Oehlke, K.; Adamiuk, M.; Behsnilian, D.; Graf, V.; Mayer-Miebach, E.; Walz, E.; Greiner, R. Potential bioavailability enhancement of bioactive compounds using food-grade engineered nanomaterials: A review of the existing evidence. *Food Funct.* **2014**, *5*, 1341–1359. [CrossRef] [PubMed]

182. Varadinova, M.G.; Docheva-Drenska, D.I.; Boyadjieva, N.I. Effects of anthocyanins on learning and memory of ovariectomized rats. *Menopause* **2009**, *16*, 345–349. [CrossRef] [PubMed]

183. Li, L.; Li, W.; Jung, S.W.; Lee, Y.W.; Kim, Y.H. Protective effects of decursin and decursinol angelate against amyloid β-protein-induced oxidative stress in the PC12 cell line: the role of Nrf2 and antioxidant enzymes. *Biosci. Biotechnol. Biochem.* **2011**, *75*, 434–442. [CrossRef] [PubMed]

184. Fomina, N.; Sankaranarayanan, J.; Almutairi, A. Photochemical mechanisms of light-triggered release from nanocarriers. *Adv. Drug Deliv. Rev.* **2012**, *64*, 1005–1020. [CrossRef] [PubMed]

185. Alvarez-Lorenzo, C.; Bromberg, L.; Concheiro, A. Light-sensitive intelligent drug delivery systems. *Photochem. Photobiol.* **2009**, *85*, 848–860. [CrossRef] [PubMed]

186. Bisby, R.H.; Mead, C.; Morgan, C.G. Wavelength-programmed solute release from photosensitive liposomes. *Biochem. Biophys. Res. Commun.* **2000**, *276*, 169–173. [CrossRef] [PubMed]

187. Morgan, C.G.; Yianni, Y.P.; Sandhu, S.S.; Mitchell, A.C. Liposome fusion and lipid exchange on ultraviolet irradiation of liposomes containing a photochromic phospholipid. *Photochem. Photobiol.* **1995**, *62*, 24–29. [CrossRef] [PubMed]

188. Panyam, J.; Labhasetwar, V. Sustained cytoplasmic delivery of drugs with intracellular receptors using biodegradable nanoparticles. *Mol. Pharm.* **2004**, *1*, 77–84. [CrossRef] [PubMed]

189. Bareford, L.M.; Swaan, P.W. Endocytic mechanisms for targeted drug delivery. *Adv. Drug Deliv. Rev.* **2007**, *59*, 748–758. [CrossRef] [PubMed]

190. Sha, X.Y.; Guo, J.; Chen, Y.Z.; Fang, X.L. Effect of phospholipid composition on pharmacokinetics and biodistribution of epirubicin liposomes. *J. Liposome Res.* **2012**, *22*, 80–88. [CrossRef] [PubMed]

191. Angelova, A.; Angelov, B.; Drechsler, M.; Lesieur, S. Neurotrophin delivery using nanotechnology. *Drug Discov. Today* **2013**, *18*, 1263–1271. [CrossRef] [PubMed]

192. Bertram, J.P.; Rauch, M.F.; Chang, K.; Lavik, E.B. Using polymer chemistry to modulate the delivery of neurotrophic factors from degradable microspheres: Delivery of BDNF. *Pharm. Res.* **2010**, *27*, 82–91. [CrossRef] [PubMed]

193. Checa-Casalengua, P.; Jiang, C.; Bravo-Osuna, I.; Tucker, B.A.; Molina-Martínez, I.T.; Young, M.J.; Herrero-Vanrell, R. Preservation of biological activity of glial cell line-derived neurotrophic factor (GDNF) after microencapsulation and sterilization by gamma irradiation. *Int. J. Pharm.* **2012**, *436*, 545–554. [CrossRef] [PubMed]

194. Garbayo, E.; Ansorena, E.; Lanciego, J.L.; Blanco-Prieto, M.J.; Aymerich, M.S. Long-term neuroprotection and neurorestoration by glial cell-derived neurotrophic factor microspheres for the treatment of Parkinson's disease. *Mov. Disord.* **2011**, *26*, 1943–1947. [CrossRef] [PubMed]

195. Lampe, K.J.; Kem, D.S.; Mahoney, M.J.; Bjugstad, K.B. the administration of BDNF and GDNF to the brain via PLGA microparticles patterned within a degradable PEG-based hydrogel: Protein distribution and the glial response. *J. Biomed. Mater. Res.* **2011**, *96*, 595–607. [CrossRef] [PubMed]

196. Huang, X.; Brazel, C.S. On the importance and mechanisms of burst release in matrix-controlled drug delivery systems. *J. Control. Release* **2001**, *73*, 121–136. [CrossRef]

197. Pridgen, E.M.; Langer, R.; Farokhzad, O.C. Biodegradable, polymeric nanoparticle delivery systems for cancer therapy. *Nanomedicine* **2007**, *2*, 669–680. [CrossRef] [PubMed]

198. Yamamoto, H.; Tahara, K.; Kawashima, Y. Nanomedical system for nucleic acid drugs created with the biodegradable nanoparticle platform. *J. Microencapsul.* **2012**, *29*, 54–62. [CrossRef] [PubMed]

199. Maurer-Jones, M.A.; Bantz, K.C.; Love, S.A.; Marquis, B.J.; Haynes, C.L. Toxicity of therapeutic nanoparticles. *Nanomedicine* **2009**, *4*, 219–241. [CrossRef] [PubMed]

200. Shvedova, A.A.; Kagan, V.E.; Fadeel, B. Close encounters of the small kind: Adverse effects of man-made materials interfacing with the nano-cosmos of biological systems. *Annu. Rev. Pharm. Toxicol.* **2010**, *50*, 63–88. [CrossRef] [PubMed]

201. Vega-Villa, K.R.; Takemoto, J.K.; Yáñez, J.A.; Remsberg, C.M.; Forrest, M.L.; Davies, N.M. Clinical toxicities of nanocarrier systems. *Adv. Drug Deliv. Rev.* **2008**, *60*, 929–938. [CrossRef] [PubMed]

202. Lanone, S.; Boczkowski, J. Biomedical applications and potential health risks of nanomaterials: Molecular mechanisms. *Curr. Mol. Med.* **2006**, *6*, 651–663. [CrossRef] [PubMed]
203. Curtis, J.; Greenberg, M.; Kester, J.; Phillips, S.; Krieger, G. Nanotechnology and nanotoxicology: A primer for clinicians. *Toxicol. Sci.* **2006**, *25*, 245–260. [CrossRef]
204. Moghimi, S.M.; Hunter, A.C.; Murray, J.C. Nanomedicine: Current status and future prospects. *FASEB J.* **2005**, *19*, 311–330. [CrossRef] [PubMed]
205. Minekus, M.; Alminger, M.; Alvito, P.; Ballance, S.; Bohn, T.; Bourlieu, C.; Carrière, F.; Boutrou, R.; Corredig, M.; Dupont, D.; et al. A standardised static in vitro digestion method suitable for food—An international consensus. *Food Funct.* **2014**, *5*, 1113–1124. [CrossRef] [PubMed]

*foods*

MDPI

*Article*

# Protective Effect of Selenium-Enriched Ricegrass Juice against Cadmium-Induced Toxicity and DNA Damage in HEK293 Kidney Cells

**Rattanamanee Chomchan [1], Sunisa Siripongvutikorn [2], Pattaravan Maliyam [3], Bandhita Saibandith [1] and Panupong Puttarak [3,4,*]**

[1]   Interdisciplinary Graduate School of Nutraceutical and Functional Food, Prince of Songkla University, Hat-Yai, Songkhla 90112, Thailand; pui_galz@hotmail.com (R.C.); bandhita.s@psu.ac.th (B.S.)
[2]   Department of Food Technology, Faculty of Agro-Industry, Prince of Songkla University, Hat-Yai, Songkhla 90112, Thailand; sunisa.s@psu.ac.th
[3]   Department of Pharmacognosy and Pharmaceutical Botany, Faculty of Pharmaceutical Sciences, Prince of Songkla University, Hat-Yai, Songkhla 90112, Thailand; Tak_kypoko@hotmail.com
[4]   Phytomedicine and Pharmaceutical Biotechnology Excellence Center (PPBEC), Prince of Songkla University, Hat-Yai, Songkhla 90112, Thailand
*   Correspondence: panupong.p@psu.ac.th; Tel.: +66-74-288-893

Received: 12 April 2018; Accepted: 24 May 2018; Published: 28 May 2018

**Abstract:** Cadmium (Cd) contamination in food is a problem endangering human health. Cd detoxication is an interesting topic particularly using food which provides no side effects. Ricegrass juice is a squeezed juice from young rice leaves which is introduced as a functional drink rich in polyphenol components. Se-enrichment into ricegrass is initiated to provide extra advantages of their functional properties. The protective role of ricegrass juice (RG) and Se-enriched ricegrass juice (Se-RG) against Cd toxicity during pre-, co- and post-treatment on HEK293 kidney cells were investigated. Results confirmed that RG and Se-RG had very low toxicity for kidney cells. Both extracts showed a protective role during pre-treatment and co-treatment against Cd toxicity by exerting a reduction in malondialdehyde (MDA) content and the percentage of DNA damage in tail and tail length of the comets over the Cd-treated cells. However, the Se-RG indicated additional benefits in all properties over RG. High Se content in Se-RG resulted in more protective effects of the regular ricegrass juice. In summary, this study provides clear evidence that Se-enriched ricegrass juice has potential to be developed as a functional food to protect the human body from Cd contamination via the reduction of oxidative stress and DNA damage.

**Keywords:** anti-cadmium toxicity; comet assay; DNA protective; HEK293 cells; ricegrass juice; selenium enrichment

## 1. Introduction

Currently, daily food consumption could lead to the unexpected exposure of contaminated compounds in the human body. Heavy metals, known as harmful agents, enter the food chain excessively via industrial operations, mining, sewage sludge and waste disposal from households into agricultural lands and water resources [1]. Accumulation of heavy metals in the environment has been reported as increasing substantially over the past decades [2]. Due to the highly soluble ability of cadmium (Cd) compounds as compared to other metals, Cd is readily taken up by plants resulting in food and feed accumulation. Cd contamination from the environment is a subject of serious health complications affecting cellular organelles and components such as cell membrane, mitochondrial, lysosome as well as genetic DNA [3]. The kidney is a critical target organ where Cd is predominantly bioaccumulated. The exposed level of Cd can cause chronic difficulties, thus leading

*Foods* **2018**, *7*, 81; doi:10.3390/foods7060081    135    www.mdpi.com/journal/foods

to damage of kidney filtering mechanisms, kidney dysfunction, liver damage as well as damage to skeletal, reproductive and respiratory systems [4,5]. The mechanism of Cd toxicity is related to its interaction with carboxyl and thiol groups of protein which generate the production of reactive oxygen species (ROS) such as superoxide ions, hydrogen peroxides, and hydroxyl radicals, and therefore inducing oxidative stress and DNA damage by initiation of lipid peroxidation [6].

Recently, several studies have been reported that antioxidant molecules have protective effects against renal and hepatic cadmium toxicity via a function of free radical scavenging [7]. Plants are the foremost source of natural antioxidant molecules such as vitamin C, vitamin E, polyphenols and some minerals like Se and Zn [8], thus they can be hypothesized as effective anti-cadmium toxicity materials. The discovery of functional plant food rich in antioxidant compounds is now being considered. Sprouts or young plants of cereals, grains or legumes are currently of interest since plants at the beginning of the growing stage are associated with large amounts of quality bioactive compounds and antioxidant molecules like polyphenols. Ricegrass is a brand-new sprout which was recently introduced as a substitute for wheatgrass, particularly in tropical areas, as a low-cost ingredient. It is rich in polyphenol compounds and has been previously investigated for its ability to scavenge free radicals in vitro effectively [9].

It was proposed that the human body attempts to reduce heavy metals toxicity via some antioxidant mechanisms such as metal chelation or degradation of free radicals [10]. Selenium (Se) has been stated as a cofactor of antioxidant enzymes and can be used as antidote agent for mercury (Hg), cadmium (Cd) and silver (Ag) [11]. The enrichment of Se into plants has been studied worldwide to increase the level of Se content and may possibly propose an extra role for the biological properties in plant foods, although concern must be given to appropriate forms and concentrations of Se [12]. Therefore, the objective of this study was to identify the specific phenolic types of polyphenol in ricegrass juice extract (RG) and Se-rich ricegrass juice extract (Se-RG) and to investigate the effect of them on in vitro anti-cadmium toxicity in HEK293 (human embryonic kidney cells kidney cells), lipid peroxidation and DNA protective properties.

## 2. Materials and Methods

### 2.1. Reagents

Di-sodium ethylenediaminetetraacetic acid (EDTA-Na$_2$), malondialdehyde (MDA), 2-thiobarbituric acid (TBA) and 3-(4,5-dimethylthiazol-2-yl)-2,5-diphenyltetrazolium bromide (MTT) were acquired from Sigma Aldrich Co. (St. Louis, MO, USA). SYBR gold nucleic acid stain and trichloroacetic acid (TCA) were purchased from Thermo Fisher Scientific Co. (San Jose, CA, USA). Reagents and media for the cell line included a trypan blue dye, trypsin-EDTA, fetal bovine serum (FBS), penicillin-streptomycin and Dulbecco's Modified Eagle Medium (DMEM) were purchased from Gibco BRL, Life Technologies Inc. (Rockville, MD, USA). Low melting point agarose (LMA), dimethyl sulfoxide (DMSO) and Triton-X was purchased from Amresco Inc. (Solon, OH, USA).

### 2.2. Plant Materials

*Oryza sativa* L. cv. Chainat 1 obtained from the Phatthalung Rice Research Center, Phatthalung, Thailand was used in this study. Se in the form of sodium selenite at range 40 mg Se/L was used to produce Se rich ricegrass as the best enrichment condition according to previous work [13]. Regular young ricegrass and Se-rich ricegrass was grown for 8 d. After harvesting, both grasses were aqueous extracted and the juices were lyophilized into powder. Ricegrass juice extract (RG) and Se-enriched ricegrass juice extract (Se-RG) were determined for the total Se content using induced coupled plasma optical emission spectroscopy (ICP-OES) and the contents were reported as 1.3 and 59.8 µg/g of extract, respectively.

## 2.3. Polyphenols Identification

RG and Se-RG were investigated for the major compounds using the reversed-phase ultra-high-performance liquid chromatography–electron spray ionization–mass detector (UHPLC–ESI–MS) since only restricted data has been stated earlier on the specific types of polyphenols found in the aqueous extract of young ricegrass. Polyphenols in RG and Se-RG were identified using a Thermo Scientific (Dionex Softron GmbH., Germering, Germany) Ultimate 3000 UHPLC system equipped with diode array absorbance detector, electron spray ionization and linear ion trap (LTQ XL) mass detector (DAD-ESI-MS). 10 mg/mL of extracts were dissolved in HPLC water and filtered with a sterile syringe filter of 0.45 μm. Purosper STAR (250 mm × 4.6 mm) with LiChrocart, Reverse Phase-18 column end-capped with 5 μm diameter particles (Merck, Darmstadt, Germany) was used as the stationary phase. $H_2O$ containing 0.5% formic acid (solvent A) and acetonitrile (solvent B) were selected as the mobile phase. The gradient condition was run according to Table 1 and followed by washing with 100% methanol for 15 min and re-equilibration. The flow rate was 0.8 mL/min, column temperature was 40 °C with the injection volume of 20 μL. The MS parameters were as follows for both the negative and positive mode: heater temperature: 250 °C; capillary temperature: 330 °C; sheath gas flow: 50 arbitrary units; auxiliary gas flow: 10 arbitrary units. The mass data for the molecular ions were processed with Thermo XcaliburTM software version 2.2.44 (Thermo Scientific., Hemel Hempstead, UK). The peaks were examined based on the ultraviolet (UV) spectra and mass ion compared to the existing literature.

**Table 1.** The gradient condition of ultra-high-performance liquid chromatography–diode array absorbance detector–electron spray ionization–mass detector (UHPLC–DAD–ESI–MS).

| Time (min) | % Solvent B |
|---|---|
| 0.00–10.00 | 0.00–10.00 |
| 10.01–15.00 | 10.00 |
| 15.01–20.00 | 10.00–15.00 |
| 20.01–30.00 | 15.00–25.00 |
| 30.01–35.00 | 25.00 |
| 35.01–45.00 | 25.00–100.00 |

## 2.4. Cell Culture Model

HEK293, human embryonic kidney cells, were purchased from the American Type Culture Collection (Manassas, VA, USA). Dulbecco's Modified Eagle Medium (DMEM) supplemented with 1.5% sodium bicarbonate, 10% fetal bovine serum (FBS), 1% penicillin-streptomycin was used for the maintenance of cells at 37 °C, 5% $CO_2$, in a fully humidified incubator. Phosphate buffer saline (PBS) at pH 7.2 was used to wash the cells through the experiment.

## 2.5. Cell Viability Assay

Functional food products which can be claimed as safe need to be confirmed on their viability effect of mammalian cells. The experiment was operated on human embryonic kidney cells, HEK293, to define the dose of extracts indicated as safe to the cells to be used for the anti-cadmium toxicity test. The MTT assay was used to determine the cell viability. Cells grown at 80–90% confluent were harvested with 0.25% trypsin–EDTA and suspended in a fresh medium. Cell counts were measured using a standard haemocytometer based trypan blue cell counting technique [14]. HEK293 cells at the density of $1 \times 10^6$ cells/mL were seeded in 96-well tissue culture plates and allowed to adhere for 24 h. After the cells were washed with PBS (pH 7.2), the media was mixed with a various concentration of the extracts (250–10,000 μg/mL), then applied to the cells followed by 24 h incubation. After 24 h, the cell viability was evaluated by adding 20 μL of MTT solution and 2 h incubation. Afterward, the MTT solution was removed, and 100 μL of 0.04 N HCl in isopropanol was added to dissolve the formazan crystals. Absorbances were recorded at 570 nm using a microplate reader. The percentage of cell viability was calculated with Equation (1);

$$\% \text{ Cell viability} = (\text{Absorbance of sample}/\text{Absorbance of control}) \times 100 \tag{1}$$

### 2.6. Anti-Cadmium Toxicity Properties

To examine the anti-cadmium toxicity properties of RG and Se-RG in HEK293 cells, initially, the half maximal cytotoxicity concentration ($CC_{50}$) of $CdCl_2$ was investigated to be used as the established dose to induce toxicity to the cells. After seeding HEK293 cells into 96-well plates at a density of $1 \times 10^6$ cells/mL, the cells were left to attach for 24 h before being treated with either $CdCl_2$ or extracts. The experiment was divided into three groups separated by different time order for treating the cells with extracts. The first treatment group was the extracts' pre-treatment; the ideal substances that can reduce toxicity from this treatment could represent the role of the protective substances. Secondly, the group of co-treatment was examined to check if the extracts could provide a protective role to cells while it directly reacted to Cd. Lastly, the post-treatment group was designed to indicate the role of the extracts as therapeutic agents. The detail of each group is briefly indicated in Table 2. The extract was fixed to have a contact time of 24 h on cells in all treatments. The percentage of cell viability was detected by MTT cytotoxicity assay and calculated as Equation (1). The morphology of the cells in each treatment was also observed and captured using a microscope.

**Table 2.** The experimental treatment group on anti-cadmium toxicity and DNA protective properties.

| | Control | | Negative Control | | Sample | |
|---|---|---|---|---|---|---|
| Time | 24 h + 24 h | | 24 h + 24 h | | 24 h + 24 h | |
| Pre-incubation | Media | Media | Media | $CdCl_2$ * | Extracts ** | $CdCl_2$ |
| Co-incubation | Media | - | Media + $CdCl_2$ | - | Extracts + $CdCl_2$ | - |
| Post-incubation | Media | Media | $CdCl_2$ | Media | $CdCl_2$ | Extracts |

* $CdCl_2$ at $CC_{50}$ level was used. ** Extracts from RG and Se-RG at 10,000 µg/mL were used.

### 2.7. Determination of Lipid Peroxidation

TBARS (thiobarbituric acid reactive substances) assay was used to determine the level of lipid peroxidation. The endogenous cellular fluid was extracted according to the modified method of Du, et al. [15]. Cells were harvested with 0.25% trypsin-EDTA and followed by centrifugation at $1000 \times g$ for 10 min. Cell pellets were washed with cold PBS until clean and re-suspended in 1 mL of cold PBS. Cells were lysed using a probe-type sonicator (Vibra-Cell, Sonics and Materials Inc., Newtown, CT, USA) by pulsing at 15 s on and 10 s off for 5 cycles on ice. The cell extracts were centrifuged at $10,000 \times g$ (4 °C) to discard the cell debris while supernatants were used for the determination of MDA content and protein levels. Protein content was examined using bovine serum albumin (BSA) as standard [16]. The modified method of Chen, et al. [17] was used to determine the MDA content. 1 mL of cellular extracts were mixed with 4 mL of 20% TCA containing 0.8% of TBA ($w/v$). The mixtures were heated at 95 °C for 60 min, then cooled in ice and centrifuged at $3000 \times g$ for 10 min. The absorbance was measured at 532 nm. The amount of MDA–TBA red complexes were compared to an external standard of MDA. The amount of TBARS was expressed as nmol MDA/mg protein.

### 2.8. DNA Protective Properties Using Comet Assay

DNA damage can be induced by exogenous agents such as heavy metals, polycyclic aromatic hydrocarbon from pollution, endogenous chemical genotoxic agents such as reactive oxygen species (ROS) and natural chemical reactions [18]. The damage to the cellular genome can generate errors in the transcription of DNA and protein translation which impair signaling and the cellular function and could result in the development of diseases [19]. The alkaline single cell gel electrophoresis assay or comet assay was used to evaluate the DNA damage [20]. The DNA protective properties of the RG and Se-RG extracts on HEK293 cells towards the exposure to $CdCl_2$ were investigated. Briefly, HEK293 cells were seeded at $1 \times 10^6$ cells/mL in 12-well plates and incubated at 37 °C for 24 h. The cells were treated with the following condition stated above (Table 2). Then, cells were harvested and fixed into slides which had been covered with 150 μL of 1.5% LMA as the first layer. After solidification, 20 μL of freshly prepared cell suspension with 180 μL of 0.5% LMA (ratio 1:10) was rapidly mixed by pipetting, and 80 μL of the mixture was loaded as the second layer. Then, 70 μL of 1% LMA was added on to the cell layer as the third layer. Once the gel was solidified, the slides were placed in a chilled lysis buffer containing 2.5 M NaCl, 100 mM EDTA, 100 mM Tris–HCl at pH 10 and 1% DMSO, 1% Triton X-100 for at least 2 h at 4 °C. The slides were then removed and placed in a comet assay tank (Model CSL-COM20, Cleaver Scientific, Rugby, Warwickshire, UK) filled with freshly prepared alkaline buffer at 4 °C (300 mM NaOH, 1 mM $Na_2$EDTA, pH $\geq$ 13) for 15 min to unwind the DNA, and electrophoresis was carried out at 25 V and 300 mA for 45 min. Afterward, the slides were rinsed with deionized water and neutralized gently with 0.4 M Tris–HCl buffer, pH 7.5 for 5 min. Finally, the slides were soaked in ethanol for 5 min and left at room temperature until they were completely dried. The cellular DNA was stained using SYBR gold nucleic acid stain in the dark for 20 min and visualized using a fluorescent microscope (Eclipse 80i, Nikon, Tokyo, Japan). The comet images (45–60 cells/slide) were captured and analyzed. The quantification of the DNA strand breaks was done using CometScore 2.0.0.38 software (Tritek Corp., Sumerduck, VA, USA). The % DNA in tail and tail length were obtained.

*2.9. Statistical Analysis*

Completely randomized design (CRD) was used throughout the study. All experimental data were presented as the mean ± standard deviation (SD) of three replications. Means were analyzed using analysis of variance (ANOVA). The significant differences among means were determined by Tukey's test ($p < 0.05$) using SPSS for Windows (SPSS Inc, Chicago, IL, USA).

## 3. Results and Discussion

*3.1. Polyphenols Identification Using Ultra High-Performance Liquid Chromatography–Electron Spray Ionization–Mass Detector (UHPLC–ESI–MS)*

Results of polyphenols identification showed that RG and Se-RG contain similar types of identified compounds. Six major compounds comprising more than 70% of relative content were tentatively identified (Table 3) where the molecular weight and electrospray ionization mass spectrometry of detected compounds were reported. Phenolic glycoside was detected and defined as 1-*O*-sinapoyl-β-D-glucose. The ion found with the *m/z* of 367 was defined as 3-*O*-feruloyl quinic acid due to the presence of a ferulic acid fragment at *m/z* 193. The largest compounds in RG were identified as a group of flavone glycosides including chrysoeriol arabinosyl arabinoside, tricin, swertisin and tricin-7-*O*-β-D-glucopyranoside which has been earlier reported as having been found in the leaves of *Oryza sativa* and *Triticum aestivum* (wheat) [21,22].

**Table 3.** Tentative identification of phenolic compounds in ricegrass juice extract (RG) and Se-rich ricegrass juice extract (Se-RG) analyzed by UHPLC–ESI–MS.

| Tentative Compounds | MW * | [M − H]⁺ (m/z) | References |
|---|---|---|---|
| Tricin | 330 | 329 | [22,23] |
| 1-O-Sinapoyl-β-D-glucose | 386 | 385 | [22] |
| 3-O-Feruloylquinic acid | 368 | 367 | [22,24] |
| Chrysoeriol arabinosyl arabinoside | 564 | 563 | [22,24] |
| Swertisin | 446 | 445 | [22] |
| Tricin-7-O-β-D-glucopyranoside | 492 | 491 | [25] |

\* MW: Molecular weight.

## 3.2. Cytotoxicity of Ricegrass Juice Extract (RG) and Se-Rich Ricegrass Juice Extract (Se-RG)

The dose of extracts indicated as safe to cells was used for the anti-cadmium toxicity test. Figure 1 revealed that the cell number of HEK293 slightly decreases while treated with both RG and Se-RG and remained constant while the concentration of the extracts was increased up to the dose of 10,000 μg/mL. Se-RG extracts displayed no significantly different effect on the reduction of cell numbers compared to the RG ($p < 0.05$). There was a minor reduction in cell number, however, the cells remained higher than 80% of the control which was indicated as the acceptable range classified as safe [26]. Although the alteration of cell morphology after treatment with the extracts has been detected, it was regarded as a result of the sensitivity of cells when they had encountered the foreign matter and attempted to adapt to the new environment. Therefore, from this result it can be assumed that both extracts had no or low toxicity to the kidney cells and the dose of both extracts at the highest concentration (10,000 μg/mL) could be used for the next experiment.

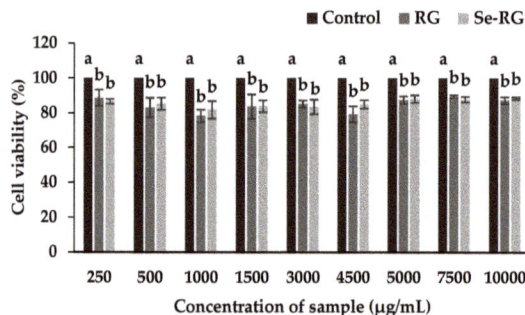

**Figure 1.** Effect of different concentrations of RG and Se-RG on the cell viability of HEK293 kidney cells. Data are means ± standard deviation (SD). Different letters indicated significant differences between treatment in the similar concentration of extracts ($p < 0.05$) in Tukey's significant differences test.

## 3.3. Anti-Cd Toxicity of RG and Se-RG and Effect on Lipid Peroxidation

The kidney is the critical organ affected by chronic Cd exposure and toxicity. Cd accumulates in the kidney because of its preferential uptake by the receptor in the renal proximal tubule and accumulates in the human kidney for a relatively long time, from 20 to 30 years [27]. The exposed level of Cd can cause chronic difficulties, thus leading to damage of kidney filtering mechanisms and kidney dysfunction [4]. HEK293 cells, human embryonic kidney cells, were used as a demonstrative model to examine the effect of $CdCl_2$ toxicity and the protective role of extracts against $CdCl_2$. The level of $CdCl_2$ which could be used to induce the cytotoxicity was examined. Figure 2 showed that the viability of HEK293 cells was significantly decreased ($p < 0.05$) while exposed to higher level of $CdCl_2$.

Cd is not able to generate radical itself but the toxicity was related to the generation of reactive oxygen species (ROS) such as superoxide ions, hydrogen peroxides and hydroxyls radicals and therefore induced oxidative stress and DNA damage by initiation of the lipid peroxidation [6].The estimated half maximal concentration (CC$_{50}$) dose of CdCl$_2$ in HEK293 was indicated as 68.5 μmol/L and this level could be used as a suitable dose for the evaluation of anti-cadmium toxicity properties of the RG and Se-RG.

The experiment on anti-cadmium toxicity properties was designed to assess the effect of RG and Se-RG against CdCl$_2$ exposure on the cell viability and lipid peroxidation at different time orders, as each substance may alleviate the toxicity from Cd induction differently [10]. Results reveal that RG and Se-RG significantly increased ($p < 0.05$) the percentage of cell viability during pre-treatment and co-treatment but not during post-treatment with Cd compared to the cells treated with Cd at CC$_{50}$ level alone (Figure 3).

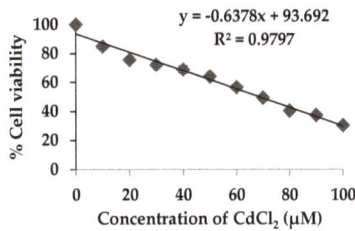

**Figure 2.** Effect of different concentrations of CdCl$_2$ on the cell viability of HEK293 kidney cells. Data are means ± standard deviation (SD).

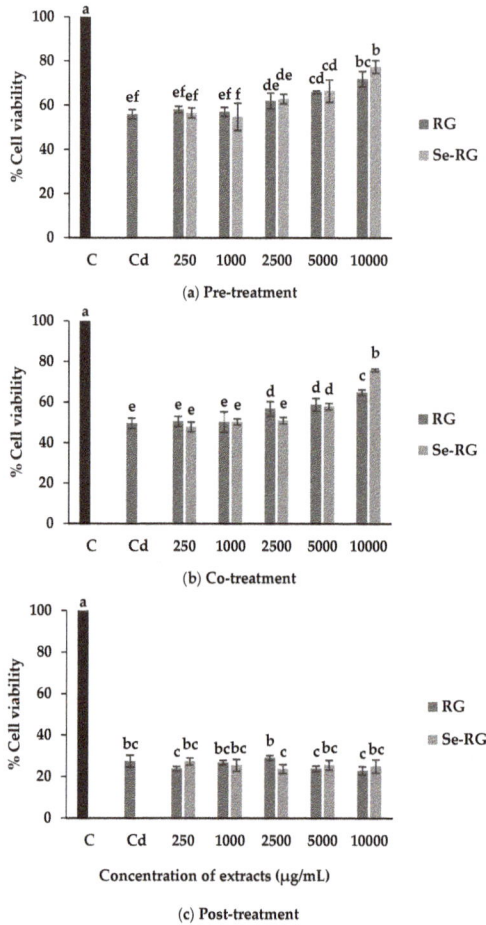

**Figure 3.** Anti-Cd toxicity properties of RG and Se-RG on HEK293 kidney cells while exposed to $CdCl_2$ at different time orders of treating the extracts compared to the control and Cd-treated (**a**) pre-incubation (**b**) co-incubation (**c**) post-incubation. C means control; Cd means $CdCl_2$ at $CC_{50}$ level (68.50 µg/mL). The cell viability was expressed as a percentage of control. Data are means ± standard deviation (SD). Different letters indicated significant differences ($p < 0.05$) in Tukey's significant differences test.

The highest concentration of both extracts (10,000 µg/mL) exerted the highest ability to protect HEK293 cells against Cd toxicity. Thus, the pathological evaluation of cell morphology treated with the extracts at this concentration was observed as shown in Figure 4. The morphological changes while treating the cells with Cd were detected. A majority of the cells were broken and floated into the media while the rest were weakened and lost their cell structure. Although the cells in the condition of extracts pre-treatment illustrated some lost and unusual cells morphology, the cells remained strengthened in their frame similarly to web shape. The changes of cells in co-treatment conditions were also detected as they were swollen and changed to a circle-like shape, but they preserved their structure. These data suggested that pre-treatment and co-treatment of RG and Se-RG with Cd could improve the Cd-induced pathological damage of kidney cells better than the Cd-treated and can potentially protect against kidney cell damage. Living organisms contain lipid as the main structure of cellular membranes. Cd could induce the damaging effects to the cells from the lipid peroxidation process [28].

Therefore, the extent of lipid peroxidation by-products produced like malondialdehyde (MDA) can imitate the extent of cells oxidative damage initiated by Cd [29]. TBARS assay is a well-established method use as an index of lipid peroxidation and lipid hydroperoxides. When the cells exposed to Cd, the MDA content was markedly increased, thus suggesting the increased in oxidative stress of kidney cells (Figure 5). However, outcomes indicated that during pre-treatment and co-treatment of both RG and Se-RG, the level of MDA in HEK293 was significantly reduced compared to Cd-treated cells ($p < 0.05$).

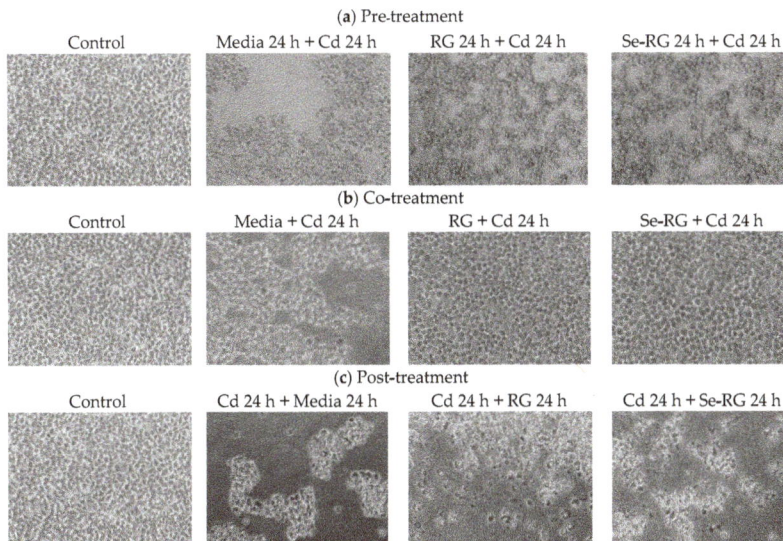

**Figure 4.** Morphology of cells of HEK293 kidney cells while incubated the cells with RG/Se-RG and CdCl$_2$ at different time order of treating the extracts (**a**) pre-incubation (**b**) co-incubation and (**c**) post-incubation. C means control; Cd means CdCl$_2$ at CC$_{50}$ level (68.50 µg/mL).

The role of phenolic compounds in the extracts was considered as having the major effects on the protective role against Cd-induced damage. It could be explained that the extracts rich in polyphenols compounds possess the inhibition of lipid peroxidation chain reaction by stabilizing the hydroxyl radicals and lipid peroxyl radicals, thereby lowering the extent of oxidative damage to the lipid cell membrane and lower level of MDA [30,31]. Moreover, phenolic compounds as antioxidant molecules could propose the role of upregulating the antioxidant protection system by stimulating the production of antioxidant enzymes including super oxide dismutase (SOD), catalase (CAT), and glutathione peroxidase (GPx). As a result, strengthening the immunity and lowering the damage caused by Cd during pre-treatment and co-treatment to the cells [10]. RG and Se-RG contained abundant polyphenols such as flavone glycosides. Therefore, the protective role of the extracts could be related mainly to these groups of compounds. Similar results also indicated the protective effect of bioflavonoids, for example, quercetin against Cd-induced oxidative stress-related renal dysfunction in rats by attenuating the Cd-induced biochemical alterations in serum, urine and tissue pathological changes via a decrease in lipid peroxidation rate [32]. Flavonoid, namely catechin from green tea, has also been proved to protect against bone metabolic disorders in cadmium-poisoned rats [33]. While focusing on the effect of high Se, Se-RG extracts revealed marginally higher protective properties against Cd toxicity over the RG in pre-treatment and co-treatment conditions. Se was used as antidote agent to a range of heavy metal toxicities including Cd, Hg, and Ag [11]. Generally, studies have indicated the beneficial effect of Se on antioxidant status and lipid peroxidation when pre-exposed and co-exposed to Cd [34–36]. Lipid peroxidation occurred because of Cd exposure; moreover, a significant decrease in the antioxidant composition factors, such as glutathione (GSH) levels, the activities of glutathione peroxidase (GPx) and thioredoxin reductase (TrxR), was also stated [34]. Se compounds have been generally known as a major cofactor of GPx and TrxR, thus, Se could logically promote the greater level of these antioxidant enzymes activity and play a role in managing the radicals occurring in the cells. Se could also present protective effects on mitochondria dysfunction by blocking the ROS generation, a possible inhibition of Cd-induced mitochondrial membrane collapse [37].

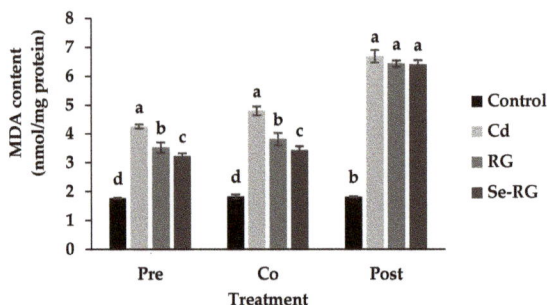

**Figure 5.** Level of malondialdehyde (MDA) in HEK293 cells treated with RG and Se-RG at the level 10,000 µg/mL compared to the control and Cd-treated ones. Data are means ± standard deviation (SD). Different letters indicated significant differences in the similar group of treatment ($p < 0.05$) in Tukey's significant differences test.

*3.4. DNA Protective Properties*

Comet assay or single cell gel electrophoresis is a standard rapid method for detecting DNA damage in individual cells. The percentage of DNA in tail and tail length was analyzed as the measure of primary DNA damage [24]. Therefore, comet assay is useful for this study to evaluate the DNA protective role of RG and Se-RG extracts against Cd. The highest concentration of both was used (10,000 µg/mL) as they protect the highest number of percent cell viability. Figure 6 illustrates the capture of comet cells of each treatment during pre-treatment, co-treatment, and post-treatment of RG and Se-RG compared to the Cd-treated and control. Figure 7 shows the parameters of the comet cells included % DNA in tail and tail length.

Comet cells of the control of every treatment displayed a circle-like shape in which the whole nuclei and DNA were beautifully stained with the fluorescent color. The condition of Cd-treated cells indicated the presence of a clouded comet tail in which the % DNA in tail and tail length were increased significantly ($p < 0.05$) because of Cd-induced the oxidative damage to cells. In both of pre-treatment and co-treatment conditions, the RG and Se-RG treated cells significantly exhibited the reduction in the % DNA in tail and tail length compared to the Cd-treated group (negative control), and thus illustrated a DNA protective effect. The results on comet assay parameters of each treatment were correlated to the content of MDA production.

The damaging of DNA could be a subsequent effect on the production of high ROS and lipid radicals induced by Cd. The role of flavone glycosides as a natural antioxidant in the RG and Se-RG may influence the protective property by possibly up-regulating the level of the antioxidant defense system and abolishing oxidative DNA damage via the donation of electrons to reactive metabolites and rendering them inactive to prevent the interaction to the DNA. The experiment on the protective role of flavonoid compounds as an excellent radical scavenger to reduce the DNA damage in human blood lymphocyte is also consistent with this result [38]. Se-RG showed higher ability on the reduction in the tail length and % DNA in the tail of the comets compared to the RG. This indicated that Se in combination with the polyphenols could provide an extra protection and promote a protective role for the kidney cells. Se as the cofactor of various endogenous enzymes works in the antioxidant system could support the activity on the destructive of ROS. Fischer, et al. [39] suggested another possible role of Se, especially in the organic form, of protecting DNA damage via induction of p53 DNA repair pathway and transactivation of p53-regulated effector genes.

In the post-treatment condition, although the addition of the RG and Se-RG showed no significant effect on an improvement in cell viability of HEK293 cells, a minor role of DNA protection can visibly be seen. In the Cd-treated condition, most DNA in cells were broken down as indicated by the size of the comet head being obviously decreased from the control. DNA fragments were spread into the surrounding area as detected from the blurred green background. However, the addition of the extracts could not save the cell viability from Cd exposure, and the intensive black background remained to be observed. This might indicate a slight reduction in the number of DNA fragments in the surrounding area as both RG and Se-RG indicated a reduction in % DNA in the tail of comets compared to Cd-treated cells. Moreover, the addition of Se-RG indicated significant reduction in the tail length ($p < 0.05$), and thus showed better DNA protective properties.

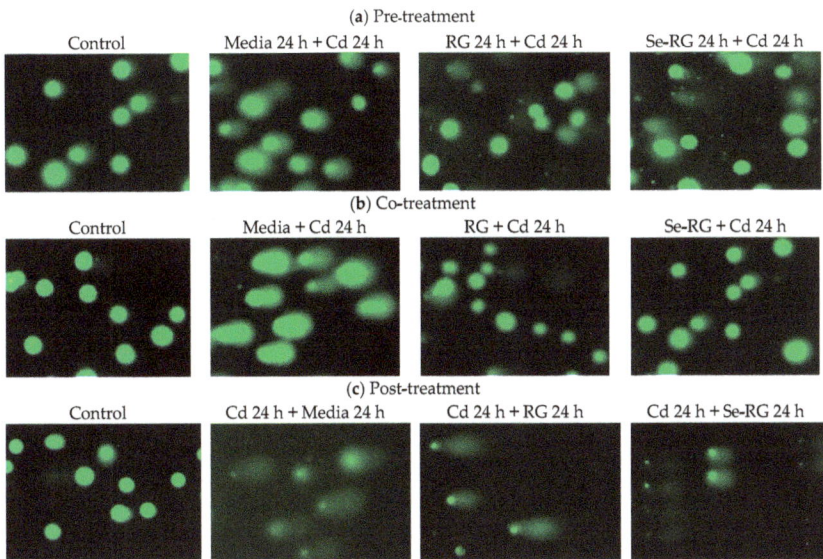

**Figure 6.** DNA protective effect of RG and Se-RG against $CdCl_2$ induced DNA break in HEK293 kidney cells while incubated the cells with RG/Se-RG and $CdCl_2$ at different time order of treating the extracts evaluated by comet assay (**a**) pre-incubation (**b**) co-incubation and (**c**) post-incubation. C means blank control; Cd means $CdCl_2$ at $CC_{50}$ level (68.50 µg/mL).

(a)

(b)

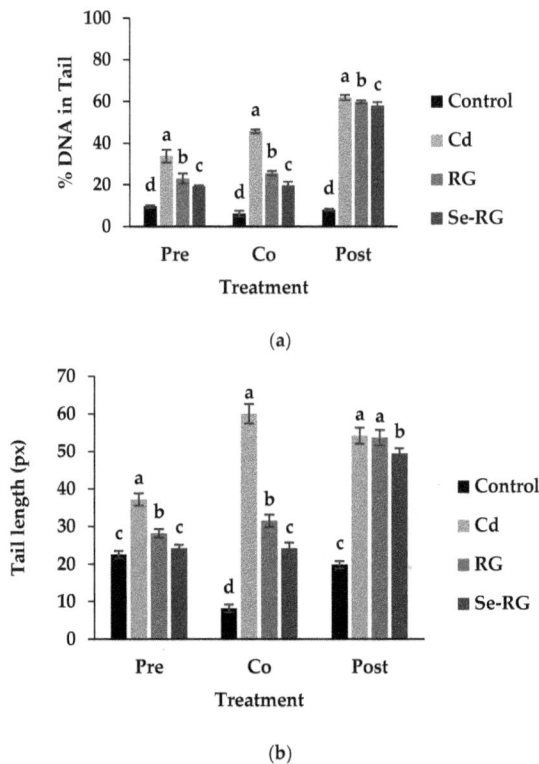

**Figure 7.** Assessment of DNA damage parameters in HEK293 cells treated with RG and Se-RG at the level 10,000 µg/mL compared to the control and Cd-treated ones evaluated using the CometScore software. (**a**) changes in % DNA in tail; (**b**) changes in tail length. Data are means ± standard deviation (SD). Different letters indicated significant differences ($p < 0.05$) in Tukey's significant differences test.

## 4. Conclusions

In conclusion, our results showed the protective effect of RG and Se-RG in counteracting Cd-induced damage in HEK293 kidney cells during pre-treatment and co-treatment with Cd. The results could support the hypothesis that polyphenols in combination with Se compounds may help in the reduction of oxidative stress, lipid peroxidation rate, morphological impairments and DNA damage to kidney cells. Flavone glycosides as the major polyphenols in the extracts should contribute to these beneficial effects. Se-enrichment to ricegrass could promote additional benefits over typical ricegrass through the upregulation of the GPx enzyme. Se-RG should have potential to be produced and consumed as a functional food to protect the human body from Cd contamination.

**Author Contributions:** R.C., P.P., and S.S. conceived and designed the experiments; R.C. performed the experiments with the assistance of P.M. and B.S.; R.C. analyzed the data and wrote the paper which was reviewed, edited and approved by all the authors.

**Funding:** This work was carried out with the financial support of Thailand's Education Hub for Southern Region of ASEAN Countries Scholarship, Graduate School Dissertation Funding for Thesis and Revenue Budget Fund (AGR600544S) from Prince of Songkla University, Thailand.

**Acknowledgments:** The authors would like to acknowledge the guidance of comet assay techniques from Pasika Niamrit, Mahidol University and English copy editing from Maria Mullet.

**Conflicts of Interest:** The authors declare no conflict of interest.

## References

1.  Tchounwou, P.B.; Yedjou, C.G.; Patlolla, A.K.; Sutton, D.J. Heavy metal toxicity and the environment. In *Molecular, Clinical and Environmental Toxicology, Experientia Supplementum*; Luch, A., Ed.; Springer: Basel, Switzerland, 2012; Volume 101, pp. 133–164. ISBN 978-3-7643-8339-8.
2.  Elinder, C.G.; Järup, L. Cadmium exposure and health risks: Recent findings. *Ambio* **1996**, *25*, 370–373.
3.  Wang, X.Q.; Xu, D.; Lü, M.K.; Yuan, D.R.; Yin, X.; Zhang, G.H.; Xu, S.X.; Lu, G.W.; Song, C.F.; Guo, S.Y. Crystal growth and physical properties of UV nonlinear optical crystal zinc cadmium thiocyanate, ZnCd(SCN)₄. *Chem. Phys. Lett.* **2001**, *346*, 393–406. [CrossRef]
4.  Järup, L. Hazards of heavy metal contamination. *Br. Med. Bull.* **2003**, *68*, 167–182. [CrossRef] [PubMed]
5.  Rahimzadeh, M.R.; Rahimzadeh, M.R.; Kazemi, S.; Moghadamnia, A.A. Cadmium toxicity and treatment: An update. *Casp. J. Intern. Med.* **2017**, *8*, 135–145.
6.  Liu, J.; Qu, W.; Kadiiska, M.B. Role of oxidative stress in cadmium toxicity and carcinogenesis. *Toxicol. Appl. Pharmacol.* **2009**, *238*, 209–214. [CrossRef] [PubMed]
7.  Shaikh, Z.A.; Vu, T.T.; Zaman, K. Oxidative stress as a mechanism of chronic cadmium-induced hepatotoxicity and renal toxicity and protection by antioxidants. *Toxicol. Appl. Pharmacol.* **1999**, *154*, 256–263. [CrossRef] [PubMed]
8.  Nahak, G.; Suar, M.; Sahu, R.K. Antioxidant Potential and Nutritional Values of Vegetables: A Review. *Res. J. Med. Plant* **2014**, *8*, 50–81. [CrossRef]
9.  Chomchan, R.; Siripongvutikorn, S.; Puttarak, P.; Rattanapon, R. Investigation of phytochemical constituents, phenolic profiles and antioxidant activities of ricegrass juice compared to wheatgrass juice. *Funct. Foods Health Dis.* **2016**, *6*, 822–835.
10. Sandbichler, A.M.; Höckner, M. Cadmium protection strategies—A hidden trade-off? *Int. J. Mol. Sci.* **2016**, *17*, 139. [CrossRef] [PubMed]
11. Mukherjee, A.; Sharma, A. Effects of cadmium and selenium on cell division and chromosomal aberrations in *Allium sativum* L. *Water Air Soil Pollut.* **1988**, *37*, 433–438. [CrossRef]
12. Chomchan, R.; Siripongvutikorn, S.; Puttarak, P. Selenium bio-fortification: An alternative to improve phytochemicals and bioactivities of plant foods. *Funct. Foods. Health Dis.* **2017**, *7*, 263–279.
13. Chomchan, R.; Siripongvutikorn, S.; Puttarak, P.; Rattanapon, R. Influence of selenium bio-fortification on nutritional compositions, bioactive compounds content and anti-oxidative properties of young ricegrass (*Oryza sativa* L.). *Funct. Food Health. Dis.* **2017**, *7*, 195–209.
14. Louis, K.S.; Siegel, A.C. Cell viability analysis using trypan blue: Manual and automated methods. In *Mammalian Cell Viability. Methods in Molecular Biology (Methods and Protocols)*; Stoddart, M., Ed.; Humana Press: New York, NY, USA, 2011; Volume 740, pp. 7–12. ISBN 978-1-61779-107-9.
15. Du, Y.; Esfandi, R.; Willmore, W.G.; Tsopmo, A. Antioxidant activity of oat proteins derived peptides in stressed hepatic HepG2 cells. *Antioxidants* **2016**, *5*, 39. [CrossRef] [PubMed]
16. Bradford, M.M. A rapid and sensitive method for the quantitation of microgram quantities of protein utilizing the principle of protein-dye binding. *Anal. Biochem.* **1976**, *72*, 248–254. [CrossRef]
17. Chen, L.; Yang, X.; Jiao, H.; Zhao, B. Tea catechins protect against lead-induced cytotoxicity, lipid peroxidation, and membrane fluidity in HepG2 cells. *Toxicol. Sci.* **2002**, *69*, 149–156. [CrossRef] [PubMed]
18. Ercal, N.; Gurer-Orhan, H.; Aykin-Burns, N. Toxic metals and oxidative stress part I: Mechanisms involved in metal-induced oxidative damage. *Curr. Top. Med. Chem.* **2001**, *1*, 529–539. [CrossRef] [PubMed]
19. Lindahl, T.; Barnes, D. Repair of endogenous DNA damage. *Cold Spring Harb. Symp. Quant. Biol.* **2000**, *65*, 127–134. [CrossRef] [PubMed]
20. Singh, N.P. Microgels for estimation of DNA strand breaks, DNA protein crosslinks and apoptosis. *Mutat. Res.* **2000**, *455*, 111–127. [CrossRef]
21. Besson, E.; Dellamonica, G.; Chopin, J.; Markham, K.R.; Kim, M.; Koh, H.-S.; Fukami, H. C-glycosylflavones from *Oryza sativa*. *Phytochemisty* **1985**, *24*, 1061–1064. [CrossRef]
22. Yang, Z.; Nakabayashi, R.; Okazaki, Y.; Mori, T.; Takamatsu, S.; Kitanaka, S.; Kikuchi, J.; Saito, K. Toward better annotation in plant metabolomics: Isolation and structure elucidation of 36 specialized metabolites from *Oryza sativa* (rice) by using MS/MS and NMR analyses. *Metabolomics* **2014**, *10*, 543–555. [CrossRef] [PubMed]

23. Kim, J.H.; Cheon, Y.M.; Kim, B.G.; Ahn, J.H. Analysis of flavonoids and characterization of the *OsFNS* gene involved in flavone biosynthesis in rice. *J. Plant Biol.* **2008**, *51*, 97–101. [CrossRef]

24. Pourrut, B.; Pinelli, E.; Celiz Mendiola, V.; Silvestre, J.; Douay, F. Recommendations for increasing alkaline comet assay reliability in plants. *Mutagenesis* **2015**, *30*, 37–43. [CrossRef] [PubMed]

25. Henkler, F.; Brinkmann, J.; Luch, A. The role of oxidative stress in carcinogenesis induced by metals and xenobiotics. *Cancers* **2010**, *2*, 376–396. [CrossRef] [PubMed]

26. Langdon, S.R.; Mulgrew, J.; Paolini, G.V.; van Hoorn, W.P. Predicting cytotoxicity from heterogeneous data sources with Bayesian learning. *J. Cheminform.* **2010**, *2*, 11. [CrossRef] [PubMed]

27. Johri, N.; Jacquillet, G.; Unwin, R. Heavy metal poisoning: The effects of cadmium on the kidney. *Biometals* **2010**, *23*, 783–792. [CrossRef] [PubMed]

28. Eneman, J.; Potts, R.; Osier, M.; Shukla, G.; Lee, C.; Chiu, J.; Hart, B. Suppressed oxidant-induced apoptosis in cadmium adapted alveolar epithelial cells and its potential involvement in cadmium carcinogenesis. *Toxicology* **2000**, *147*, 215–228. [CrossRef]

29. Ayala, A.; Muñoz, M.F.; Argüelles, S. Lipid peroxidation: Production, metabolism, and signaling mechanisms of malondialdehyde and 4-hydroxy-2-nonenal. *Oxid. Med. Cell. Longev.* **2014**, *2014*, 1–31. [CrossRef] [PubMed]

30. Cai, Q.; Rahn, R.O.; Zhang, R. Dietary flavonoids, quercetin, luteolin and genistein, reduce oxidative DNA damage and lipid peroxidation and quench free radicals. *Cancer Lett.* **1997**, *119*, 99–107. [CrossRef]

31. Ding, Y.; Wang, S.Y.; Yang, D.J.; Chang, M.H.; Chen, Y.C. Alleviative effects of litchi (*Litchi chinensis Sonn.*) flower on lipid peroxidation and protein degradation in emulsified pork meatballs. *J. Food Drug Anal.* **2015**, *23*, 501–508. [CrossRef] [PubMed]

32. Renugadevi, J.; Prabu, S.M. Quercetin protects against oxidative stress-related renal dysfunction by cadmium in rats. *Exp. Toxicol. Pathol.* **2010**, *62*, 471–481. [CrossRef] [PubMed]

33. Choi, J.H.; Rhee, I.K.; Park, K.Y.; Park, K.Y.; Kim, J.K.; Rhee, S.J. Action of green tea catechin on bone metabolic disorder in chronic cadmium-poisoned rats. *Life Sci.* **2003**, *73*, 1479–1489. [CrossRef]

34. El-Sharaky, A.; Newairy, A.; Badreldeen, M.; Eweda, S.; Sheweita, S. Protective role of selenium against renal toxicity induced by cadmium in rats. *Toxicology* **2007**, *235*, 185–193. [CrossRef] [PubMed]

35. Liu, L.; Yang, B.; Cheng, Y.; Lin, H. Ameliorative effects of selenium on cadmium-induced oxidative stress and endoplasmic reticulum stress in the chicken kidney. *Biol. Trace Elem. Res.* **2015**, *167*, 308–319. [CrossRef] [PubMed]

36. Ognjanovic, B.; Markovic, S.; Pavlovic, S.; Zikic, R.; Stajn, A.; Saicic, Z. Effect of chronic cadmium exposure on antioxidant defense system in some tissues of rats: Protective effect of selenium. *Physiol. Res.* **2008**, *57*, 403–411. [PubMed]

37. Zhou, Y.J.; Zhang, S.P.; Liu, C.W.; Cai, Y.Q. The protection of selenium on ROS mediated-apoptosis by mitochondria dysfunction in cadmium-induced LLC-PK1 cells. *Toxicol. In Vitro* **2009**, *23*, 288–294. [CrossRef] [PubMed]

38. Devipriya, N.; Sudheer, A.R.; Srinivasan, M.; Menon, V.P. Quercetin ameliorates gamma radiation-induced DNA damage and biochemical changes in human peripheral blood lymphocytes. *Mutat. Res.* **2008**, *654*, 1–7. [CrossRef] [PubMed]

39. Fischer, J.L.; Lancia, J.K.; Mathur, A.; Smith, M.L. Selenium protection from DNA damage involves a Ref1/p53/Brca1 protein complex. *Anticancer Res.* **2006**, *26*, 899–904. [PubMed]

*foods*

MDPI

*Article*

# Identification of Rice *Koji* Extract Components that Increase β-Glucocerebrosidase Levels in Human Epidermal Keratinocytes

**Kazuhisa Maeda** [1,2,*], **Yuuka Ogino** [2], **Ayano Nakamura** [3], **Keiji Nakata** [3], **Manabu Kitagawa** [3] **and Seiki Ito** [3]

[1]  Bionics Program, Tokyo University of Technology Graduate School, 1404-1 Katakuramachi, Hachioji City, Tokyo 192-0982, Japan
[2]  School of Bioscience and Biotechnology, Tokyo University of Technology, 1404-1 Katakuramachi, Hachioji City, Tokyo 192-0982, Japan; yk.lax6054@gmail.com
[3]  MARUKOME Co., Ltd., 883 Amori, Nagano-shi, Nagano 380-0943, Japan; ayano_nakamura@marukome.co.jp (A.N.); keiji_nakata@marukome.co.jp (K.N.); manabu_kitagawa@marukome.co.jp (M.K.); seiki_itou@marukome.co.jp (S.I.)
*  Correspondence: kmaeda@stf.teu.ac.jp; Tel.: +81-426372442

Received: 21 April 2018; Accepted: 15 June 2018; Published: 18 June 2018

**Abstract:** Rice miso contains many ingredients derived from rice *koji* and has been a valuable source of nutrition since ancient times. We found that the consumption of rice miso led to improvements in the moisture content of cheek stratum corneum, skin viscoelasticity, and skin texture. Further, rice miso extract was found to increase the mRNA expression and activity of β-glucocerebrosidase (β-GCase), an enzyme involved in ceramide synthesis in the stratum corneum, in cultures. In this study, we identified the lipid-derived components of rice *koji* that increase the β-GCase activity in cultured human epidermal keratinocytes. The methanol fraction of rice *koji* extract induced an increase in the mRNA expression and activity of β-GCase in keratinocytes. The active fraction of rice *koji* was found to contain phosphatidic acid (PA) and lysophosphatidic acid (LPA). The total PA concentration in rice *koji* was 973.9 ng/mg dry weight, which was 17.5 times higher than that in steamed rice. Among the molecular species, PA_18:2/18:2 was the most frequently found. The total LPA concentration in rice *koji* was 29.6 ng/mg dry weight, and 2-LPA_18:2 was the most frequently found LPA. Since PA and LPA increase the mRNA expression and activity of β-GCase in keratinocytes, they are thought to be the active ingredients in rice *koji* that increase the β-GCase levels in human epidermal keratinocytes.

**Keywords:** *Aspergillus oryzae*; rice *koji*; phosphatidic acid; lysophosphatidic acid; β-glucocerebrosidase

## 1. Introduction

Rice *koji*, which is a solid-state culture of *koji* molds (*Aspergilus oryzae*) in rice, is used as a raw material for rice miso. Rice miso contains many ingredients derived from rice *koji* and has been a valuable source of nutrition since ancient times. We found that ingesting rice miso improved the moisture content in cheek stratum corneum, skin viscoelasticity, and texture [1]. Furthermore, the effect of miso extract on ceramide synthesis in epidermal keratinocytes was examined in cultured cells, and the mRNA expression and activity of glucocerebrosidase (β-GCase) in these cells increased when the cells were treated with miso extract [1]. From these results, it was hypothesized that the moisture level of the skin improves by everyday consumption of miso soup made from rice miso and that this may occur via increasing the ceramide content of the stratum corneum [1]. *A. oryzae* produces and secretes many types of enzymes, including α-amylase, which degrades starch; glucoamylase;

transglucosidase; acid protease, which degrades protein; and acidic carboxypeptidase [2,3], as it grows. In addition, rice *koji* contains various nutrients and supplies the necessary nutrients for yeast and lactic acid bacteria. These nutrients are especially rich in vitamins of the B group and include molybdenum; biotin; vitamins B1, B2, and B6; folic acid; pantothenic acid; and niacin [4]. Furthermore, mono- and diacylglycerol are converted to free fatty acids by the lipid-decomposing enzymes of rice *koji* [5], and other studies have suggested that the free fatty acid content in miso contributes to its antimutagenic activities [6]. The main constituent fatty acids of *A. oryzae* are lignoceric acid and, particularly, 2-oxy lignoceric acid and 2,3-dioxyl lignoceric acid, with phytosphingosine reported as the major long-chain constituent [7]. Rice bran malt contains a bifidobacterial growth-promoting substance and glucosylceramide, which improve intestinal bacterial flora in mice [8,9]. It has also been reported that *A. oryzae* exhibits antioxidant activity [10] and produces multiple cathepsin B-inhibiting components [11]. Rice *koji* is used for most Japanese fermented foods, including miso, soy sauce, amazake, Japanese sake, and malt vinegar. Rice *koji* and sake lees have also been used as cosmetics and skin moisturizing agents for many years. In this paper, we report on the lipid-derived components of rice *koji* that increase β-GCase levels in cultured human epidermal keratinocytes.

## 2. Materials and Methods

### 2.1. Materials

Human epidermal keratinocytes used for a three-dimensional epidermis model were purchased from the Japan Tissue Engineering Co. Ltd. (Tokyo, Japan). Rice *koji*, which is a solid-state culture of *koji* molds (*A. oryzae*) in rice, was obtained from Marukome Co., Ltd. (Nagano, Japan). Reagents not described in the text were obtained from Wako Pure Chemical Industries, Ltd. (Osaka, Japan).

### 2.2. Preparation of Rice Koji Extract

Five hundred grams of fermented rice *koji* produced by an *A. oryzae* addition to steamed rice was extracted using 3 L of chloroform/methanol (2:1). The extract was filtered and concentrated under reduced pressure on a rotary evaporator. The concentrated extract was fractionated sequentially with 3 L of chloroform, 3 L of acetone, and 3 L of methanol using a silica gel column (5 cm × 35 cm), and the fractionated solution was recovered. After concentrating this solution under reduced pressure with an evaporator, the weight of the sample was measured, and each sample was adjusted to a 5 mg/mL concentration by the addition of ethanol.

### 2.3. Effects of Fractions of Rice Koji Extracts on Cell Proliferation and β-GCase Activity in Cultured Human Keratinocytes

Human epidermal keratinocytes were seeded in a 96-well plate at a density of $1 \times 10^4$ cells/well, cultured in Dulbecco's modified Eagle's medium containing 10% fetal bovine serum (FBS-DMEM) for 24 h and exchanged into 99 μL of 2% FBS-DMEM. Subsequently, 1 μL each of the chloroform fraction, acetone fraction, and methanol fraction of the chloroform/methanol (2:1) extract of rice *koji* was added ($n = 4$). Then, the cells were cultured in a $CO_2$ incubator for three days. After culturing, 10 μL of Cell Counting Kit-8 solution (Dojindo Laboratories, Inc., Kumamoto, Japan) was added to each well, and after 2 h, the absorbance (at 450 nm) was measured with a microplate reader (Multi-Detection Microplate Powerscan HT; BioTek, Winooski, VT, USA) to obtain the number of cells. In this method, the tetrazolium salt WST-8 (2-(2-methoxy-4-nitrophenyl)-3-(4-nitrophenyl)-5-(2,4-disulfophenyl)-2H-tetrazolium, monosodium salt) is reduced by intracellular dehydrogenases, and the absorbance at 450 nm of the generated water-soluble formazan is measured to determine the number of living cells. The number of cells and the amount of formazan produced are linearly proportional.

β-GCase activity was determined using 4-methylumbelliferyl β-D-glucopyranoside [1]. Human keratinocytes cultured for three days in the 96-well plate were washed twice with PBS, and

30 μL of 0.01 mol/L acetate buffer (pH = 5.0) and 40 μL of a 5 mmol/L 4-methylumbelliferyl β-D-glucopyranoside solution were added. After a 1 h reaction in an incubator, 50 μL of a 0.1 mol/L glycine sodium hydrate buffer solution (pH = 10.7) was added to determine the fluorescence using the microplate reader. The measurement was performed at an excitation wavelength of 360 nm and an emission wavelength of 460 nm.

## 2.4. Effects of Fractions from Silica Gel Column Re-Fractionated Rice Koji Extracts on Cell Proliferation and β-GCase Activity in Cultured Human Keratinocytes

Methanol fractions ①–⑦ of rice *koji* extract were separately dried and then dissolved in 6 mL of chloroform. The entire volume of dissolved lipid solution was added to 1 g of silica gel in a column (Strata SI-1, Phenomenex Inc., Torrance, CA, USA) equilibrated with chloroform and then eluted with 40 mL of chloroform/methanol (2:1). Subsequently, phospholipids other than lysophospholipids were eluted with 40 mL of chloroform/methanol (1:1). Each fraction was weighed after being dried under reduced pressure. Then, lysophospholipids were obtained with 40 mL of methanol. Each phospholipid or lysophospholipid fraction was separated and weighed after being dried under reduced pressure. Their concentrations were adjusted to 5 mg/mL. Human keratinocytes with these fractions at a concentration of 50 μg/mL were cultured in 2% FBS-DMEM for three days, and the number of cells and β-GCase activity were measured in the same manner as in Section 2.3.

## 2.5. TLC Analysis of Components in Rice Koji That Affect β-GCase Activity

Methanol fractions ②–⑥ (chloroform/methanol 1:1 fractionation, 10 mg/mL); methanol fractions ④–⑥ (methanol re-fractionation, 10 mg/mL); 1,2-di-(cis-9-octadecenoyl)-*sn*-glycerin 3-phosphate sodium salt (PA_C18:1/C18:1, Sigma-Aldrich Corp., St. Louis, MO, USA; 10 mg/mL), as a standard of phosphatidic acid (PA); and oleoyl-L-α-lysophosphatidic acid sodium salt (1-LPA_C18:1, Sigma-Aldrich Corp., St. Louis, MO, USA; 10 mg/mL), as a standard of lysophosphatidic acid (LPA), were spotted on a thin-layer chromatography (TLC) plate (silica gel 60 TLC aluminum plates, Merck KGaA, Darmstadt, Germany). The plate was developed with chloroform/methanol/water/triethylamine (30:35:7:35 ($v/v$)). The developed plate was sprayed with molybdenum blue reagent and left at room temperature till turning a blue color. Phospholipids in the sample were estimated from the Rf value of the obtained spots. The spot density was analyzed by ImageJ. The method of preparing the molybdenum blue reagent (coloring reagent) was as follows. Solution A was prepared by adding 4.01 g of molybdenum trioxide ($MoO_3$) to 100 mL of 25 mol/L $H_2SO_4$, which was then boiled gently until the solid dissolved. Solution B was prepared by adding 0.18 g of powdered molybdenum (Mo) to 50 mL of solution A, which was then boiled gently for 15 min and allowed to cool. Equal amounts of the supernatants were mixed. After diluting three times with water, an equal volume of ethanol was added, and the solution was placed in a nebulizer.

## 2.6. Effects of PA_C18:1/C18:1, PA_C16:0/C18:1, and 1-LPA_C18:1 on Cell Proliferation and β-GCase Activity in Cultured Human Keratinocytes

PA_C18:1/C18:1,1-palmitoyl-2-oleoyl-*sn*-glycero-3-phosphatidic acid (PA_C16:0/C18:1) from the Cayman Chemical Company, Ann Arbor, MI, USA, and 1-LPA_C18:1 were used. Their concentrations were adjusted to 10 mg/mL, 1 mg/mL, 0.1 mg/mL, and 0.01 mg/mL with ethanol, respectively. The effect of PA_C18:1/C18:1, PA_C16:0/C18:1, and 1-LPA_C18:1 on the number of cells and β-GCase activity in cultured human keratinocytes was examined as described in Section 2.3.

## 2.7. Effects of the PA and LPA Fractions of Rice Koji on TEWL on a Three-Dimensional Human Epidermis Model

An agar plate was placed on a warmer set at 32 °C in a room with a temperature of 22 °C and a humidity of 40%, and a three-dimensional human epidermis model, which had been cultured for four days in the presence or absence of the PA fraction ⑤ or LPA fraction ⑤ of rice *koji* (10 μg/mL,

100 μg/mL in a solvent of ethanol:water (1:99)), was placed on the agar plate inlay. The transepidermal water loss (TEWL) was measured with a Tewameter TM 210 (Courage and Khazaka Electronic GmbH, Köln, Germany) using a probe for 24-well culture inserts.

### 2.8. Effects of PA and LPA on TEWL in a Three-Dimensional Human Epidermis Model

An agar plate was placed on a warmer set at 32 °C in a room with a temperature of 22 °C and a humidity of 40%, and a three-dimensional human epidermis model, which had been cultured for four days in the presence or absence of PA_C18:1/C18:1, PA_C16:0/C18:1, or 1-LPA_C18:1 (10 μ/mL, 100 μg/mL in a solvent of ethanol:water (1:99)), was placed on the agar plate inlay. The TEWL was measured with the Tewameter TM210 using the probe for 24-well culture inserts [12].

### 2.9. Effects of the PA and LPA Fractions of Rice Koji, PA_C18:1/C18:1, PA_C16:0/C18:1, and 1-LPA_C18:1 on the Amount of β-GCase mRNA in the Three-Dimensional Human Epidermis Model

We investigated the amount of β-GCase mRNA involved in ceramide synthesis using a human epidermal model cultured for four days in the absence or presence of the PA fraction of rice *koji*, the LPA fraction, PA_C18:1/C18:1, PA_C16:0/C18:1, or 1-LPA_C18:1 (10 μg/mL, 100 μg/mL in a solvent of ethanol:water (1:99)). RNA extraction was performed using the RNeasy Protect Mini Kit (RNeasy Mini kit, Qiagen GmbH, Hilden, Germany) according to the manufacturer's instructions. cDNA conversion from the extracted RNA was performed using the One Step SYBR® Prime Script™ RT-PCR Kit II (Takara Bio Inc., Shiga, Japan) according to the package insert. The expression level of mRNA was measured using real-time PCR (Applied Biosystems 7900HT, Thermo Fisher Scientific Inc., Waltham, MA, USA). GAPDH was used as a reference gene for normalization. Primers were purchased from Qiagen. Each experiment was performed thrice. In the analysis, a threshold cycle Ct was obtained, the difference between the Ct value of the housekeeping gene and the Ct value of the target gene (ΔCt: (target gene Ct) − (housekeeping gene Ct)) was obtained, and the ratio of gene expression was calculated from the difference in ΔCt between samples (ΔΔCt).

### 2.10. Measurement of the PA and LPA Contents in Steamed Rice and Rice Koji

Five hundred microliters of methanol and 500 μL of chloroform were added to 20 mg of lyophilized powder of steamed rice and rice *koji*, and after shaking and centrifugation, 100 μL of supernatant was collected. Then, 100 μL of a methanol solution of a PA measurement internal standard (PA_16:0 D 31/18: 1 (500 ng/mL)) was added to this supernatant sample to yield an analytical sample for PA measurement. In addition, 300 μL of the supernatant was taken, and 150 μL of the methanol solution of internal standard 1-LPA_17:0, 2-LPA_17:0 (50 ng/mL) was added to prepare an analytical sample for LPA measurement. Liquid chromatography (LC) and mass spectrometry (MS) conditions are shown below.

LC conditions

| | |
|---|---|
| Device | UltiMate 3000 BioRS (Thermo Fisher Scientific Inc.) |
| Column | L-column 2 ODS metal-free column (2 mm ID × 150 mm, 3 μm, Chemicals Evaluation and Research Institute, Tokyo, Japan) |
| Flow rate | 0.2 mL/min |
| Column temperature | 40 °C |
| Sampler temperature | 5 °C |
| Injection volume | 3 μL |

MS conditions

| | |
|---|---|
| Device | 3200 QTRAP (AB Sciex Pte Ltd. Framingham, MA, USA) |
| Ionization method | Electrospray ionization |
| Measurement mode | SRM (negative) |

The analysis software Analyst 1.6.1 (AB Sciex) was used for peak detection. The detection limit was set to a signal noise ratio (S/N) of 3. The area value of the detected peak was divided by the peak area value of the internal standard (PA_16:0 D 31/18:1) to calculate the ratio. The approximate concentration of each molecular species in the sample was calculated using the obtained peak area ratio. The coefficient of variation (C.V. %) for five measurements of the peak area ratio (analysis sample/internal standard) of each molecular species was $6.9 \pm 2.5$ (mean $\pm$ standard deviation).

*2.11. Measurement of the Relative LPA Contents in Steamed Rice and Rice Koji*

To prepare the lysophospholipid fraction, 15 mL of chloroform/methanol (2:1) was added to 5 g of each of the three types of steamed rice and rice *koji*, and the mixtures were stirred and extracted at room temperature for 1 h. They were divided into soluble and solid components by filtration, and distilled water was added while stirring to make a chloroform/methanol/water solution (8:4:3 volume ratio). The sample was left undisturbed, and the lower layer (chloroform layer, total lipid fraction) was collected. The total amount of collected lipids was added to a silica gel column (Strata SI-1 100 mg (Phenomenex Inc.), gel volume 1 mL) equilibrated with chloroform; neutral lipids were then eluted using 10 mL of chloroform, and glycolipids were eluted using 40 mL of acetone. Finally, a total phospholipid fraction was obtained with 10 mL of methanol. Total phospholipids were dried and then dissolved in 1 mL of chloroform. The dissolved phospholipid solution was added in its entirety to a silica gel column (Strata SI-1 100 mg, Phenomenex Inc.) equilibrated with chloroform, and phospholipids other than lysophospholipids were eluted with 10 mL of chloroform/methanol (1:1). Subsequently, a lysophospholipid fraction was obtained with 10 mL of methanol. This fraction was dried under reduced pressure and dissolved in 100 μL of 10% Triton X-100-isopropanol. For a determination of LPA concentration, the lysophospholipid fraction was methyl-esterified using a fatty acid methyl esterification kit (Nacalai Tesque, Inc., Kyoto, Japan), and the products were analyzed by GC-flame ionization detection (FID) (TC-WAX, GL Sciences Inc., Tokyo, Japan). 1-LPA_C18:1 (Tocris Bioscience, Bristol, UK) was used as a standard. The peak area ratio of each molecular species with a C.V. below 15% for three measurements was adopted.

*2.12. Statistical Analysis*

Statistical analysis was performed using t-tests run with MS Excel. Differences with a *p* value less than 5% were statistically significant (* $p < 0.05$, ** $p < 0.01$, *** $p < 0.001$), and there was no significant difference (ns) above 5%. In addition, when the ratio was greater than 5% but less than 10%, it was considered to be exhibiting a tendency ($^{+}$ $p < 0.1$).

## 3. Results

*3.1. Effects of Rice Koji Extract Fractions on Cell Proliferation and β-GCase Activity in Cultured Human Keratinocytes*

Figure 1 shows the effects of the chloroform, acetone, and methanol fractions of rice *koji* extract on the number of cells and β-GCase activity in cultured human keratinocytes. Significantly more cells were observed in treatments with methanol fractions ②–⑦ of the rice *koji* extract than in the control (Figure 1a). In addition, a significant enhancement in β-GCase activity was confirmed in cells treated with methanol fractions ⑤ and ⑥ of the rice *koji* extract (Figure 1b). The chloroform and acetone fractions of rice *koji* extract had no such effect.

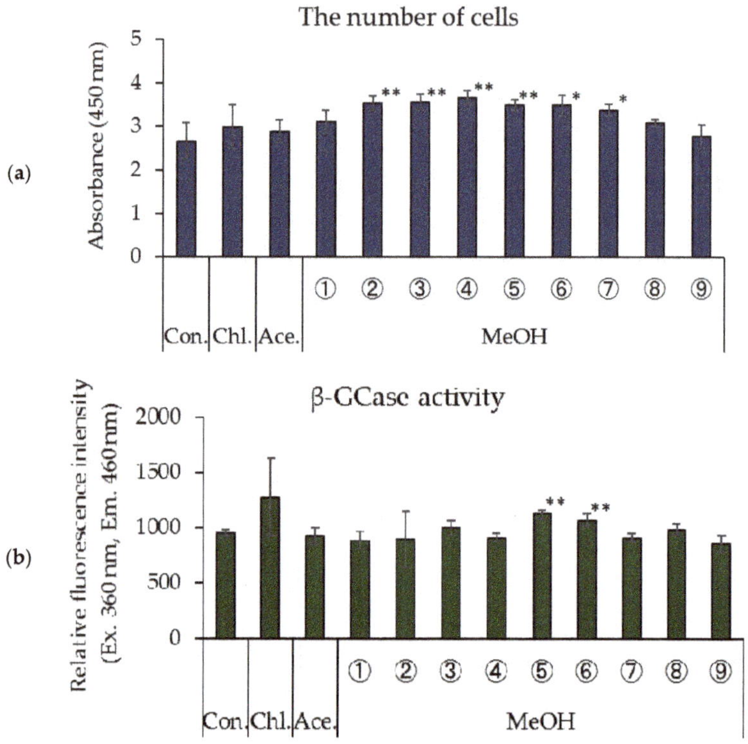

**Figure 1.** Effects of fractions of rice *koji* extracts on cell proliferation and β-GCase activity in cultured human keratinocytes. (**a**) The number of cells, (**b**) β-GCase activity, $n = 4$, mean ± standard deviation (S.D.), $^+ p < 0.1$ vs. control, $^* p < 0.05$ vs. control, and $^{**} p < 0.01$ vs. control.

### 3.2. Effects of Fractions from Silica Gel Column Re-Fractionated Rice Koji Extracts on Cell Proliferation and β-GCase Activity in Cultured Human Keratinocytes

Figure 2 shows the effect of silica gel column re-fractionated rice *koji* extracts on the number of cells and β-GCase activity in cultured human keratinocytes. A significant increase in the number of cells was confirmed in treatments with silica column fractions ②, ③, and ⑤ (chloroform/methanol 1:1 re-fractionation) of the methanol fractions of rice *koji* extract, and a tendency to promote cell proliferation was observed in fraction ① (Figure 2a). In addition, a significant increase in the number of cells was observed in treatments with silica gel column fraction ① (methanol re-fractionation) of the methanol fractions of rice *koji* extract (Figure 2a). A significant increase in β-GCase activity was observed in cells treated with fractions ②–⑦ of the silica gel column fractionations (chloroform/methanol 1:1 re-fractionation) from the methanol fractions of rice *koji* extract (Figure 2b). In addition, a significant enhancement in β-GCase activity was confirmed in cells treated with silica gel column fractions ②–⑥ (methanol re-fractionation) of the methanol fractions of rice *koji* extract, and a tendency to promote β-GCase activity was observed in fraction ⑦ (Figure 2b).

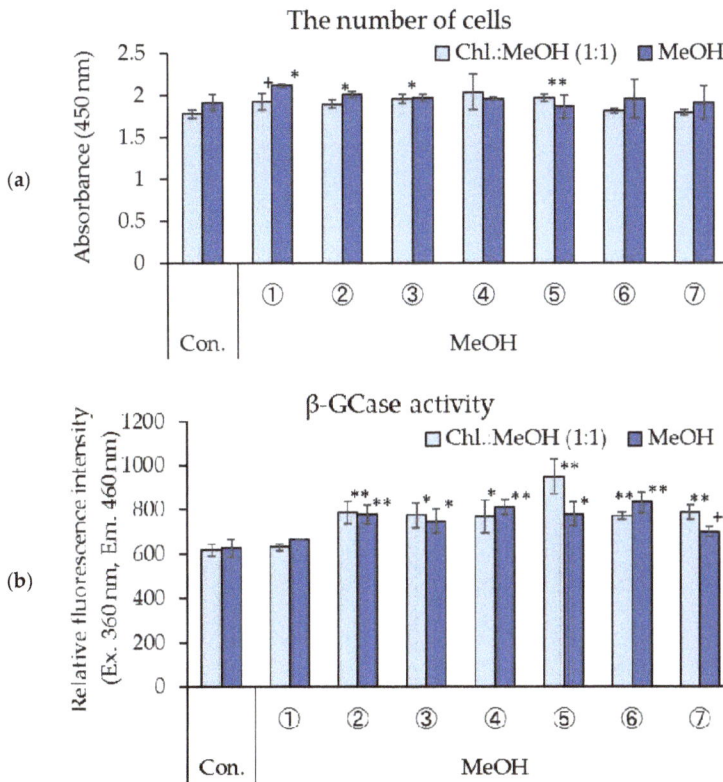

**Figure 2.** Effects of fractions from silica gel column re-fractionated rice *koji* extract on cell proliferation and β-GCase activity in cultured human keratinocytes. (**a**) The number of cells, (**b**) β-GCase activity, $n = 4$, mean ± S.D., $^{+} p < 0.1$ vs. control, $^{*} p < 0.05$ vs. control, and $^{**} p < 0.01$ vs. control.

### 3.3. TLC Analysis of Components of Rice Koji Extract That Induce β-GCase Activity

In Figure 3a, the TLC images of the chloroform and acetone extracts, methanol fractions ②–⑦ (chloroform/methanol 1:1 re-fractionation of rice *koji* extract) and methanol fractions ②–⑦ (methanol re-fractionation of rice *koji* extract) are shown. The presence of PA in methanol fractions ②–⑥ (chloroform/methanol 1:1 re-fractionation of rice *koji* extract) and LPA in methanol fractions ④–⑥ (methanol re-fractionation of rice *koji* extract) was confirmed. As a result of the analysis of spot density with ImageJ, the presence of PA in methanol fraction ⑤ (chloroform/methanol 1:1 re-fractionation of rice *koji* extract) and of LPA in methanol fractions ④ and ⑤ (methanol refraction of rice *koji* extract) was confirmed in a majority of cases (Figure 3b).

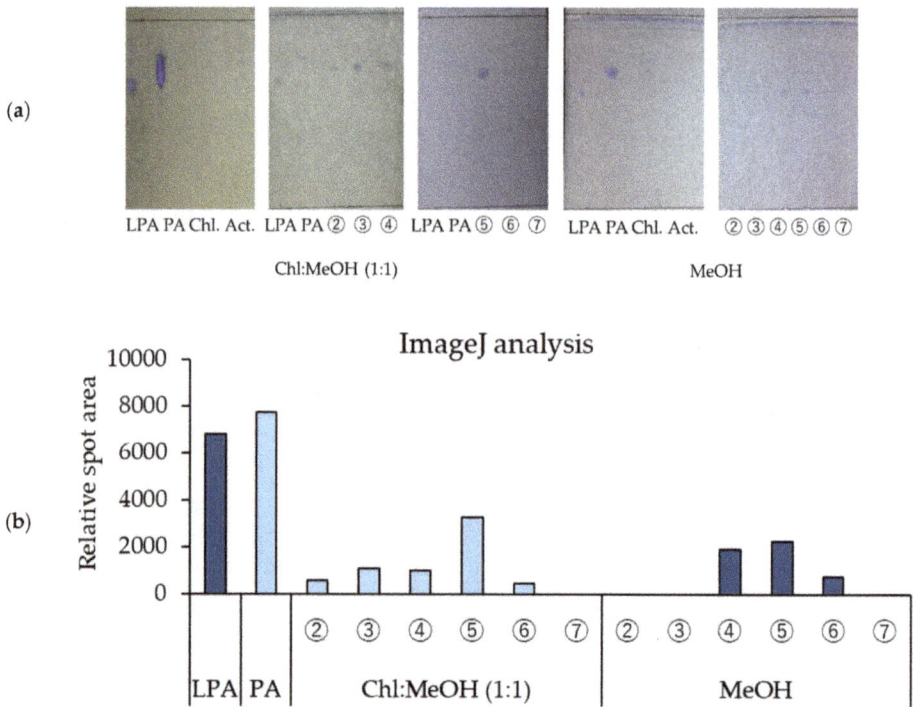

**Figure 3.** TLC (thin-layer chromatography) analysis of components in rice *koji* extract that increase β-GCase activity. (**a**) TLC image, (**b**) analysis of spot density with ImageJ. LPA: 1-LPA_C18:1, LA: PA_C18:1/C18:1.

*3.4. Effects of PA_C18:1/C18:1 and 1-LPA_C18:1 on Cell Proliferation and β-GCase Activity in Cultured Human Keratinocytes*

Figure 4 shows the effects of PA_C18:1/C18:1 and 1-LPA_C18:1 on the number of cells and β-GCase activity in cultured human keratinocytes. Treatment with PA_C18:1/C18:1 (10 μg/mL, 100 μg/mL) and 1-LPA_C18:1 (100 μg/mL) resulted in a significant increase in the number of cells (Figure 4a). In addition, a significant enhancement in β-GCase activity was observed in cells treated with either PA_C18:1/C18:1 (10 μg/mL, 100 μg/mL) or 1-LPA_C18:1 (10 μg/mL, 100 μg/mL). Treatment with PA_C18:1/C18:1 (1 μg/mL) also demonstrated a tendency towards increased β-GCase activity (Figure 4b).

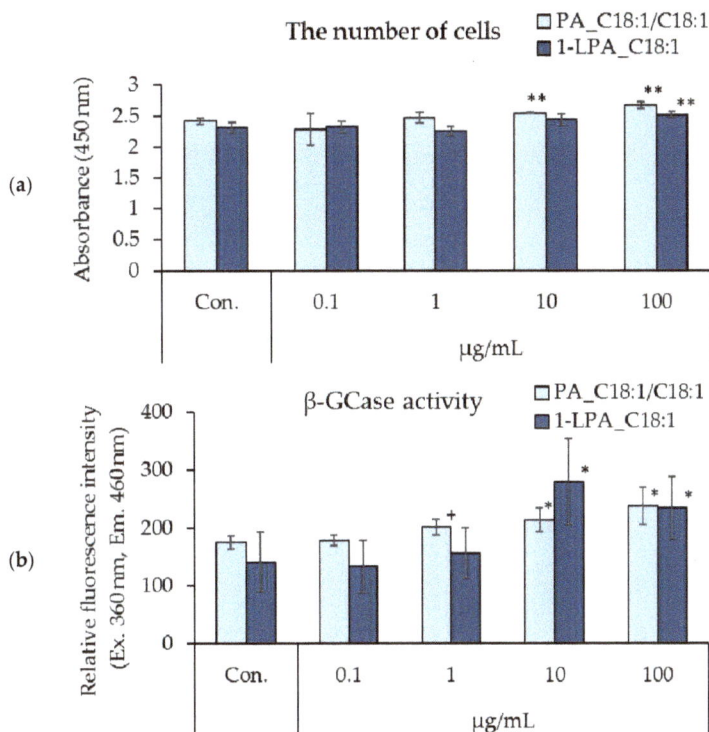

**Figure 4.** Effects of PA_C18:1/C18:1 and 1-LPA_C18:1 on cell proliferation and β-GCase activity in cultured human keratinocytes. (**a**) Number of cells, (**b**) β-GCase activity; $n = 4$, mean $\pm$ S.D., [+] $p < 0.1$ vs. control, [*] $p < 0.05$ vs. control, and [**] $p < 0.01$ vs. control.

### 3.5. Effects of the PA and LPA Fractions of Rice Koji Extract, PA_C18:1/C18:1, PA_C16:0/C18:1, and 1-LPA_C18:1 on the TEWL in the Three-Dimensional Human Epidermal Model

Figure 5a shows the TEWL, which is the amount of moisture that transpired from the stratum corneum in the three-dimensional human epidermal model cultured for four days in the presence or absence of the PA and LPA fractions of rice *koji* extract. The TEWL from human epidermal models cultured in the presence of methanol fraction ⑤ (chloroform/methanol 1:1 re-fractionation, 10 μg/mL, 100 μg/mL) as the PA fraction and in the presence of methanol fraction ⑤ (methanol re-fractionation, 100 μg/mL) as the LPA fraction was significantly lower than that in the absence of these fractions (control). In addition, the TEWL from human epidermal models cultured in the presence of the LPA fraction (10 μg/mL) had a tendency to be lower than that in the control (Figure 5a).

Furthermore, Figure 5b shows the TEWL from human epidermal models cultured for four days in the presence or absence of either PA or LPA. The TEWL from human epidermal models cultured in the presence of PA_C18:1/C18:1 (10 μg/mL, 100 μg/mL), PA_C16:0/C18:1 (10 μg/mL, 100 μg/mL), and 1-LPA_C18:1 (100 μg/mL) was significantly lower than that in the absence of these fractions (control), and the TEWL in the epidermis model cultured for four days in the presence of 1-LPA_C18:1 (10 μg/mL) showed a tendency to be lower than the TEWL in the control (Figure 5b).

The transpiration of moisture from the stratum corneum was significantly suppressed by PA and LPA.

β-GCase mRNA / G3PDH mRNA

(a)

Transepidermal water loss (TEWL)

(b)

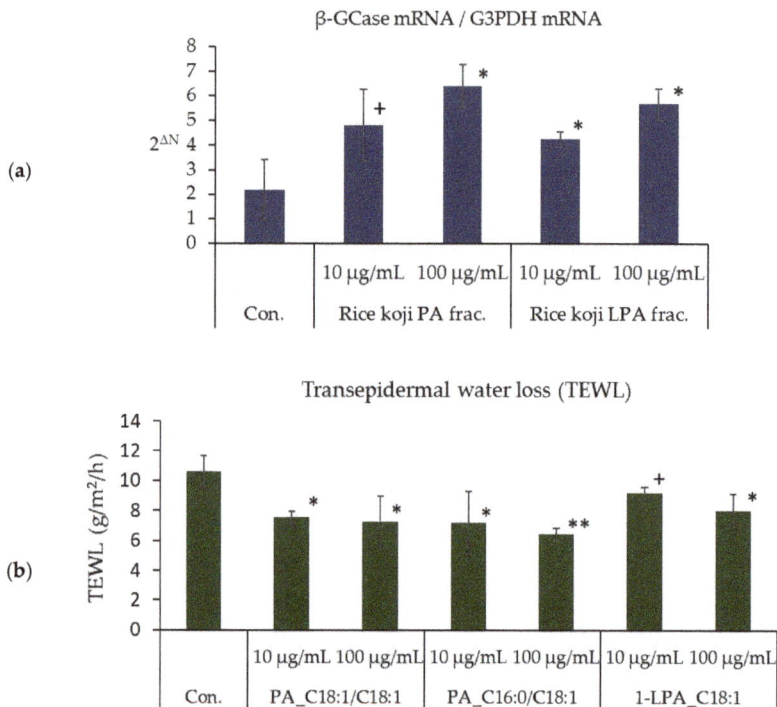

**Figure 5.** Effects of the PA and LPA fractions of rice *koji* extract, PA_C18:1/C18:1, PA_C16:0/C18:1, and 1-LPA_C18:1 on the TEWL from the three-dimensional epidermal model. (**a**) Effects of the PA and LPA fractions of rice *koji* extract on the TEWL, (**b**) effects of PA_C18:1/C18:1, PA_C16:0/C18:1, and 1-LPA_C18:1 on the TEWL; $n = 3$, mean $\pm$ S.D., $^+$ $p < 0.1$ vs. control, $^*$ $p < 0.05$ vs. control, and $^{**}$ $p < 0.01$ vs. control.

*3.6. Effects of the PA and LPA Fractions of Rice Koji Extract, PA_C18:1/C18:1, PA_C16:0/C18:1, and 1-LPA_C18:1 on the Expression Level of β-GCase mRNA in the Three-Dimensional Human Epidermis Model*

The level of β-GCase mRNA in the three-dimensional human epidermis models cultured for four days in the presence of either rice *koji* methanol fraction ⑤ (chloroform/methanol 1:1 re-fractionation, PA fraction 100 μg/mL) or methanol fraction ⑤ (methanol re-fractionation, LPA fraction 10 μg/mL, 100 μg/mL) was significantly higher than that in the control (Figure 6a). The mRNA level of β-GCase in the epidermal model cultured for four days in the presence of methanol fraction ⑤ (chloroform/methanol 1:1 re-fractionation, PA fraction 10 μg/mL) tended to be higher than the mRNA level of β-GCase in the control (Figure 6a). Furthermore, the level of β-GCase mRNA in the epidermal models cultured for four days in the presence of PA_C18:1/C18:1 (100 μg/mL), PA_C16:0/C18:1 (100 μg/mL), or 1-LPA_C18:1 (100 μg/mL) was significantly higher than that found in the control epidermal model (Figure 6b). The level of β-GCase mRNA in the models cultured for four days in the presence of PA_C18:1/C18:1 (10 μg/mL), PA_C16:0/C18:1 (10 μg/mL), or 1-LPA_C18:1 (10 μg/mL) tended to be higher than that of the control model (Figure 6b).

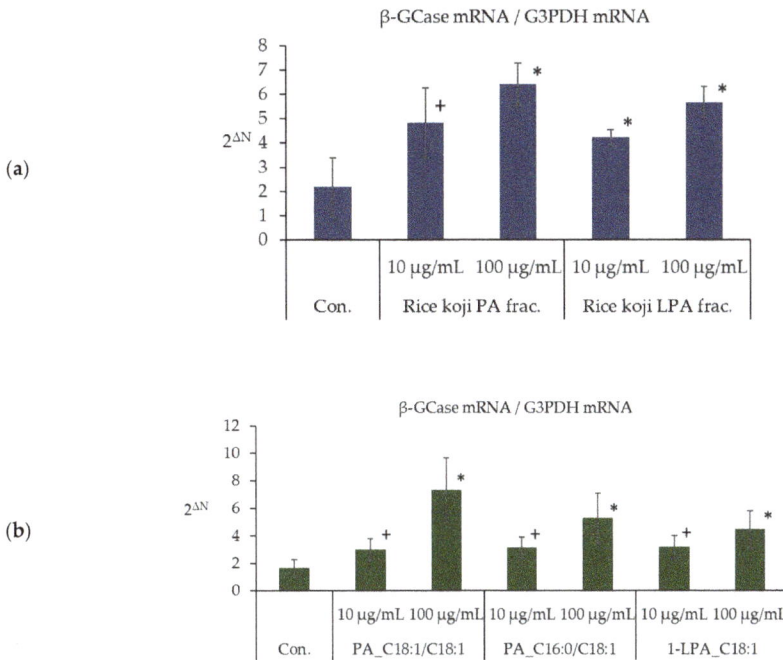

**Figure 6.** Effects of the PA and LPA fractions of rice *koji* extract, PA_C18:1/C18:1, PA_C16:0/C18:1, and 1-LPA_C18:1 on β-GCase mRNA expression in cultured human keratinocytes. (**a**) Effects of the PA and LPA fractions of rice *koji* extract on the level of β-GCase mRNA, (**b**) effects of PA_C18:1/C18:1, PA_C16:0/C18:1, and 1-LPA_C18:1 on the level of β-GCase mRNA; $n = 3$, mean ± S.D., [+] $p < 0.1$ vs. control, [*] $p < 0.05$ vs. control, and [**] $p < 0.01$ vs. control.

### 3.7. Measurement of the PA and LPA Contents in Steamed Rice and Rice Koji

The content of PA in each molecular species of steamed rice and rice *koji* was measured, and the results are shown in Figure 7a. The total PA in rice *koji* was 973.9 ng/mg dry weight, which was 17.5 times higher than the total PA in steamed rice (55.5 ng/mg of dry weight). PA_18:2/18:2 is the most common molecular species, followed by, in decreasing order, PA_16:0/18:2, PA_18: 1/18:2, and PA _16:0/18:1 (Figure 7a). The LPA contents of steamed rice and rice *koji* are shown in Figure 7b. LPA was not detected in steamed rice, but LPA at a concentration of 29.6 ng/mg dry weight was confirmed in rice *koji*. In addition, the 2-LPA content in rice *koji* was 14 times that of 1-LPA. The 2 -LPA_18:2 species was the most frequently found 2-LPA, followed by 2-LPA_18:1. Little 1-LPA was found.

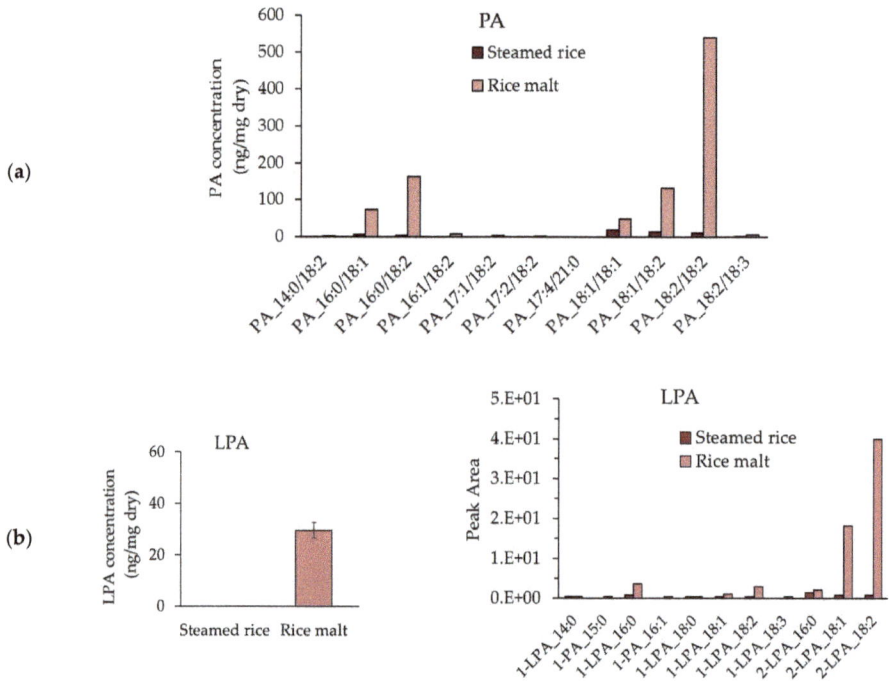

**Figure 7.** PA and LPA contents in steamed rice and rice *koji*. (**a**) Molecular species content of PA in steamed rice and rice *koji*, (**b**) LPA contents in steamed rice and rice *koji* (left), and molecular species content of LPA in rice *koji* relative to that in steamed rice (right).

*3.8. Measurement of the PC and LPC Contents in Steamed Rice and Rice Koji*

The contents of phosphatidylcholine (PC) and lysophosphatidylcholine (LPC) in steamed rice and rice *koji* are shown in Figure 8. Although PC was hardly detected in steamed rice, PC_18:2/18:2 was the most abundant PC in rice *koji*, followed by PC_18:1/18:2 and PC_16:0/18:2 (Figure 8a). Two types of 1-LPC were present in steamed rice: 1-LPC_16:0 and 1–LPC_18:2. Meanwhile, 1-LPC was also present in rice *koji*, and 1-LPC_16:0 and 1-LPC_18:2 were abundant. Both 2-LPC_16:0 and 2-LPC_18:2 were less abundant in rice *koji* than the other types of LPC (Figure 8b).

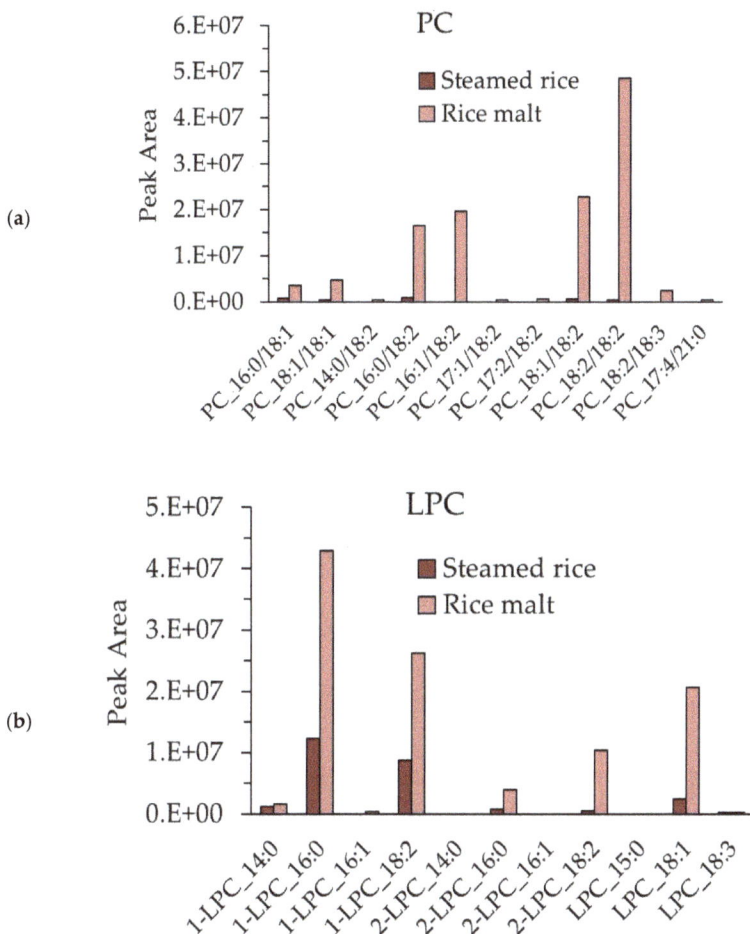

**Figure 8.** PC and LPC contents in steamed rice and rice *koji*. (**a**) Relative molecular species content of PC in steamed rice and rice *koji* and (**b**) relative molecular species content of LPC in steamed rice and rice *koji*.

## 4. Discussion

The stratum corneum intercellular lipids, which extend into the intercellular space of the stratum corneum (extracellular stratum corneum lipids), act as a barrier between the epidermis and the air of the outside world and are critical to preventing the transpiration of moisture from inside the body and moisture ingression from outside the body. When this barrier is damaged, excessive evaporation occurs, and the stratum corneum becomes dehydrated, while further deterioration causes allergic reactions in the skin, such as rough skin and atopic dermatitis [13–15]. The intercellular lipids consist of ceramides (37%), cholesterols (32%), long-chain fatty acids (16%), and cholesterol esters (15%) and form a multilayered lamellar lipid structure [16]. Among these components, ceramides are particularly important in the stratum corneum and are present in amounts 30 times or greater in the stratum corneum than in other organs. The stratum corneum ceramides are characterized not only by their large quantity, but also by the presence of various kinds of molecules [17,18]. Acylceramide, which is important for skin barrier function, is produced by β-GCase from acylglucosylceramide [19,20]. In addition, some ceramides are produced from both glucosylceramide and sphingomyelin [21,22].

We showed that the mRNA expression and activity of β-GCase, which is an enzyme involved in ceramide synthesis in the stratum corneum, increases with exposure to rice miso extract [1]. The levels of phytosphingosine- and sphingosine-type ceramides increase after such exposure [1]. During the fermentation of miso, various functional substances are produced by the enzymatic action of various fermenting microorganisms. For example, miso contains functional substances, such as essential fatty acids (linoleic acid and linolenic acid), and acylphosphatidic acid is produced by *Rhizopus oryzae* [23]. In this study, to investigate whether ceramide synthesis is involved in the skin improvement effects of rice *koji*, we examined the effects of the lipid component of rice *koji* on the mRNA level and activity of β-GCase in cultured keratinocytes. We found that both the mRNA level and activity of β-GCase were increased in cells treated with a methanol fraction of rice *koji* extract and with PA and LPA. In addition, when a methanol fraction of rice *koji* extract, PA, and LPA were allowed to act on the three-dimensional human epidermal skin model, a decrease in TEWL and increases in β-GCase activity and β-GCase mRNA expression occurred. On the other hand, the chloroform and acetone fractions of rice *koji* extract had no such effect. Rice (*Oryza sativa* L.) contains phospholipase A1 and phospholipase A2 [24], and rice bran oil contains phospholipase D [25,26]. Since steamed rice is used for the production of rice *koji*, these enzymes are thought to be inactivated. On the other hand, *A. oryzae* produces phospholipase A1 [27,28]. Rice bran oil contains phospholipids [29], and rice contains many PCs, which are phospholipids, and LPC_16:0 (3.0 to 4.7 µg/mL) and LPC_18:2 (0.8 to 2.2 µg/mL), which are both lysophospholipids [30].

In our study, PA and LPA were included in rice *koji*, and since their concentrations were higher in rice *koji* than in steamed rice, the source of phospholipids was thought to be the rice. The abundance of 2-LPA was 14 times greater than that of 1-LPA in rice *koji*. The 2-LPA_18:2 species was the most frequently found, followed by 2-LPA_18:1, and the 1-LPA level was low in rice *koji*. From these results, PA was considered to be produced from PC by phospholipase D of *A. oryzae*, and 2-LPA was produced from PA by phospholipase A1. Furthermore, these results indicated that the activity of phospholipase A1 of *A. oryzae* was higher than that of phospholipase A2. However, since the PA content in rice *koji* was 10 times greater than the LPA content, the phospholipase A1 activity of *A. oryzae* was not thought to be high. In addition, 2-LPA may be produced by phospholipase D of *A. oryzae* from the 2-LC produced by phospholipase A1 of *A. oryzae*, but the pathway by which 1-LPA is produced from 1-LC via phospholipase D of *A. oryzae* contributes less. The activity of phospholipase D of orthologous genes has not been verified in *A. oryzae* [31], and further investigating the variations in phospholipase D for each *A. oryzae* strain is necessary.

## 5. Conclusions

The methanol fraction of rice *koji* extract contains PA and LPA, both of which increase the mRNA expression and activity of β-GCase in cultured human epidermal keratinocytes. Therefore, PA and LPA are thought to be among the active ingredients in rice *koji* that increase the β-GCase levels in human epidermal keratinocytes. The total PA in rice *koji* was 973.9 ng/mg dry weight, which was 17.5 times higher than that in the steamed rice used as the source of rice *koji*, and contained several molecular species, of which PA_18:2/18:2 was the most abundant. The total LPA in rice *koji* was 29.6 ng/mg dry weight, and 2-LPA_18:2 was the most abundant. Based on these findings, foods containing rice *koji* (which includes high levels of PA and LPA) can presumably function in improving the moisture content, viscoelasticity, and texture of human skin.

**Author Contributions:** K.M. conceived and designed the experiments; Y.O. performed the experiments; K.M. and Y.O. analyzed the data; K.N., A.N., M.K., and S.I. were engaged in the manufacture of the rice *koji*; and K.M. wrote the paper.

**Funding:** This research received no external funding.

**Acknowledgments:** The study was supported by a grant from the Nagano Prefecture General Industrial Technology Center. Measurements of the PA and LPA contents in steamed rice and rice *koji* were performed by the Kansai Research Institute, Inc., Japan and the Chemicals Evaluation and Research Institute, Japan.

**Conflicts of Interest:** The authors declare no conflicts of interest.

## References

1. Maeda, K.; Nakata, K.; Nakamura, A.; Kitagawa, M.; Ito, S. Improvement in skin conditions by consumption of traditional Japanese miso soup and its mechanism. *J. Nutr. Food Sci.* **2018**, *8*, 1. [CrossRef]
2. Kitamoto, K. *Bunshi Kojikin Gaku*; Revised Version; The Brewing Society of Japan: Tokyo, Japan, 2012. (In Japanese)
3. Iwano, K. Seisyu shuzou to kouso. *J. Brew. Soc. Jpn.* **1974**, *74*, 206–212. (In Japanese)
4. Ministry of Education, Culture, Sports, Science and Technology, Japan. Standard Tables of Food Composition in Japan (Seventh Revised Version). 2015. Available online: http://www.mext.go.jp/en/policy/science_technology/policy/title01/detail01/1374030.htm (accessed on 14 April 2018).
5. Ohnishi, K.; Yoshida, Y.; Toita, J.; Sekiguchi, J. Purification and characterization of a novel lipolytic enzyme from *Aspergillus oryzae*. *J. Ferment. Bioeng.* **1994**, *78*, 413–419. [CrossRef]
6. Watanabe, T.; Owari, K.; Hori, K.; Takahashi, K. Selection of *koji* mold strain for making functional miso as rich antimutagenic activity. *Nippon Shokuhin Kagaku Kogaku Kaishi* **2004**, *51*, 698–702. (In Japanese) [CrossRef]
7. Ohnishi, M.; Fujino, Y. Molecular species of free ceramides in *Aspergillus oryzae*. *Agric. Biol. Chem.* **1976**, *40*, 1419–1426. [CrossRef]
8. Hosoyama, H.; Oisawa, M.; Hamano, M. Bifidobacterium growth promoting substance in rice bran *koji* extract. *Nippon Shokuhin Kogyo Gakkaishi* **1991**, *38*, 940–944. [CrossRef]
9. Hamajima, H.; Matsunaga, H.; Fujikawa, A.; Sato, T.; Mitsutake, S.; Yanagita, T.; Nagao, K.; Nakayama, J.; Kitagaki, H. Japanese traditional dietary fungus *koji Aspergillus oryzae* functions as a prebiotic for *Blautia coccoides* through glycosylceramide: Japanese dietary fungus *koji* is a new prebiotic. *Springerplus* **2016**, *5*, 1321. [CrossRef] [PubMed]
10. Kawasumi, T.; Sato, M.; Tsuriya, Y.; Ueno, S. Antioxidant activity of a medicine based on *Aspergillus oryzae* NK *koji* measured by a modified t-butyl peroxyl radical scavenging assay. *Biosci. Biotechnol. Biochem.* **1999**, *63*, 581–584. [CrossRef] [PubMed]
11. Yamada, T.; Hiratake, J.; Aikawa, M.; Suizu, T.; Saito, Y.; Kawato, A.; Suginami, K.; Oda, J. Cysteine protease inhibitors produced by the industrial *koji* mold, *Aspergillus oryzae* O-1018. *Biosci. Biotechnol. Biochem.* **1998**, *62*, 907–914. [CrossRef] [PubMed]
12. Morimoto, H.; Gu, L.; Zeng, H.; Maeda, K. Amino carbonylation of epidermal basement membrane inhibits epidermal cell function and is suppressed by methylparaben. *Cosmetics* **2017**, *4*, 38. [CrossRef]
13. Feingold, K.R.; Elias, P.M. Role of lipids in the formation and maintenance of the cutaneous permeability barrier. *Biochim. Biophys. Acta.* **2014**, *1841*, 280–294. [CrossRef] [PubMed]
14. Groen, D.; Poole, D.S.; Gooris, G.S.; Bouwstra, J.A. Investigating the barrier function of skin lipid models with varying compositions. *Eur. J. Pharm. Biopharm.* **2011**, *79*, 334–342. [CrossRef] [PubMed]
15. Elias, P.M.; Menon, G.K. Structural and lipid biochemical correlates of the epidermal permeability barrier. *Adv. Lipid Res.* **1991**, *24*, 1–26. [PubMed]
16. Norlén, L.; Nicander, I.; Lundh Rozell, B.; Ollmar, S.; Forslind, B. Inter- and intra- individual differences in human stratum corneum lipid content related to physical parameters of skin barrier function *in vivo. J. Invest. Dermatol.* **1999**, *112*, 72–77. [CrossRef] [PubMed]
17. Hamanaka, S.; Hara, M.; Nishio, H.; Otsuka, F.; Suzuki, A.; Uchida, Y. Human epidermal glucosylceramides are major precursors of stratum corneum ceramides. *J. Invest. Dermatol.* **2002**, *119*, 416–423. [CrossRef] [PubMed]
18. Van Smeden, J.; Hoppel, L.; van der Heijden, R.; Hankemeier, T.; Vreeken, R.J.; Bouwstra, J.A. LC/MS analysis of stratum corneum lipids: Ceramide profiling and discovery. *J. Lipid Res.* **2011**, *52*, 1211–1221. [CrossRef] [PubMed]
19. Holleran, W.M.; Takagi, Y.; Menon, G.K.; Jackson, S.M.; Lee, J.M.; Feingold, K.R.; Elias, P.M. Permeability barrier requirements regulate epidermal beta-glucocerebrosidase. *J. Lipid Res.* **1994**, *35*, 905–912. [PubMed]

20. Hanley, K.; Jiang, Y.; Holleran, W.M.; Elias, P.M.; Williams, M.L.; Feingold, K.R. Glucosylceramide metabolism is regulated during normal and hormonally stimulated epidermal barrier development in the rat. *J. Lipid Res.* **1997**, *38*, 576–584. [PubMed]

21. Uchida, Y.; Hara, M.; Nishio, H.; Sidransky, E.; Inoue, S.; Otsuka, F.; Suzuki, A.; Elias, P.M.; Holleran, W.M.; Hamanaka, S. Epidermal sphingomyelins are precursors for selected stratum corneum ceramides. *J. Lipid Res.* **2000**, *41*, 2071–2082. [PubMed]

22. Jensen, J.M.; Fölster-Holst, R.; Baranowsky, A.; Schunck, M.; Winoto-Morbach, S.; Neumann, C.; Schütze, S.; Proksch, E. Impaired sphingomyelinase activity and epidermal differentiation in atopic dermatitis. *J. Invest. Dermatol.* **2004**, *122*, 1423–1431. [CrossRef] [PubMed]

23. Oliveira Mdos, S.; Feddern, V.; Kupski, L.; Cipolatti, E.P.; Badiale-Furlong, E.; de Souza-Soares, L.A. Changes in lipid, fatty acids and phospholipids composition of whole rice bran after solid-state fungal fermentation. *Bioresour. Technol.* **2011**, *102*, 8335–8338. [CrossRef] [PubMed]

24. Singh, A.; Baranwal, V.; Shankar, A.; Kanwar, P.; Ranjan, R.; Yadav, S.; Pandey, A.; Kapoor, S.; Pandey, G.K. Rice phospholipase A superfamily: Organization, phylogenetic and expression analysis during abiotic stresses and development. *PLoS ONE* **2012**, *7*, e30947. [CrossRef] [PubMed]

25. Lee, M.H. Phospholipase D of rice bran. I. Purification and characterization. *Plant Sci.* **1989**, *59*, 25–33. [CrossRef]

26. Ueki, J.; Morioka, S.; Komari, T.; Kumashiro, T. Purification and characterization of phospholipase D (PLD) from rice (*Oryza sativa* L.) and cloning of cDNA for PLD from rice and maize (*Zea mays* L.). *Plant Cell Physiol.* **1995**, *36*, 903–914. [CrossRef] [PubMed]

27. Watanabe, I.; Koishi, R.; Yao, Y.; Tsuji, T.; Serizawa, N. Molecular cloning and expression of the gene encoding a phospholipase A1 from *Aspergillus oryzae*. *Biosci. Biotechnol. Biochem.* **1999**, *63*, 820–826. [CrossRef] [PubMed]

28. Shiba, Y.; Ono, C.; Fukui, F.; Watanabe, I.; Serizawa, N.; Gomi, K.; Yoshikawa, H. High-level secretory production of phospholipaseA1 by *Saccharomyces cerevisiae* and *Aspergillus oryzae*. *Biosci. Biotechnol. Biochem.* **2001**, *65*, 94–101.

29. Liu, L.; Waters, D.L.; Rose, T.J.; Bao, J.; King, G.J. Phospholipids in rice: significance in grain quality and health benefits: A review. *Food Chem.* **2013**, *139*, 1133–1145. [CrossRef] [PubMed]

30. Tong, C.; Liu, L.; Waters, D.L.; Rose, T.J.; Bao, J.; King, G.J. Genotypic variation in lysophospholipids of milled rice. *J. Agric. Food Chem.* **2014**, *62*, 9353–9361. [CrossRef] [PubMed]

31. AspGD. *A. oryzae* AO090005000433 Summary. The Board of Trustees, Leland Stanford Junior University. Available online: http://www.aspergillusgenome.org/cgi-bin/locus.pl?locus=AO090005000433#citations (accessed on 14 April 2018).

MDPI

St. Alban-Anlage 66

4052 Basel

Switzerland

Tel. +41 61 683 77 34

Fax +41 61 302 89 18

www.mdpi.com

*Foods* Editorial Office

E-mail: foods@mdpi.com

www.mdpi.com/journal/foods

www.ingramcontent.com/pod-product-compliance
Lightning Source LLC
Chambersburg PA
CBHW051900210326

41597CB00033B/5968